Pregnancy at Work

The Maternity Alliance

working to make life better for pregnant women, new parents and their babies

The Maternity Alliance is a national charity which aims to improve the care, health, education and social support given to parents before conception, during pregnancy, childbirth and in the first year of their child's life. We are the only organisation which directly challenges maternity poverty having been founded in 1980 by organisations concerned by the clear link between poverty and maternity and infant mortality and morbidity. We therefore focus on pregnant women and new parents on low incomes, those with disabilities and those from ethnic minorities, believing that services which get things right for them will result in improvements for everyone.

Regina H. Kenen
with Jenny McLeish and Daphne May

Pregnancy at Work

Health and Safety for the Working Woman

Pluto Press
LONDON • CHICAGO, ILLINOIS
in association with Maternity Alliance

First published 1998 by Pluto Press
345 Archway Road, London N6 5AA
and 1436 West Randolph, Chicago, Illinois 60607, USA

British Library Cataloguing in Publication Data
A catalogue record for this book is available from
the British Library

ISBN 0 7453 1291 8 hbk

Library of Congress Cataloging in Publication Data

Designed and produced for Pluto Press by
Chase Production Services, Chadlington, Oxford, OX7 3LN
Typeset from disk by Stanford DTP Services, Northampton
Printed in the EC by TJ International, Padstow

To Joanne, Marc, Stephanie and Judith, my own 'reproductive hazards', and to my husband Peter who helped create them.

Contents

Preface

The number of pregnant women and mothers in the paid work force with children under 16 has burgeoned. Between 1984 and 1994 the percentage of women in the UK who worked outside the home and had a dependent child (or children) increased from 55 per cent to 64 per cent, with the likelihood of such women being employed increasing as the youngest child got older. Despite this second fact, the last decade has seen the greatest increase among mothers whose offspring were five years old or younger: from 37 per cent in 1984 to 52 per cent in 1994.

These women are more aware and concerned about occupational and environmental hazards than they have ever been. They have heard both accurate and inaccurate reports about possible harmful effects of asbestos insulation crumbling in ceilings, vapours emitted by photocopying machines, anaesthetic gases leaking in operating rooms, hair dyes and cosmetic preparations used in beauty salons, and chemical plant spills. They understand that physical, chemical and biological substances as well as psychological and social pressures can harm them when they are pregnant or can make their pregnancies more difficult. They have heard that workplace hazards may cause miscarriages, stillbirths, and an increase in the number of children born with physical abnormalities, mental retardation and learning disabilities. Furthermore, children of exposed workers may face a higher risk of developing childhood cancer. Some of these reproductive threats, such as exposure to lead, are well documented. Others are strongly suspected. Still others, such as the effects of working long hours at video display terminals, have been reported by women, but are still under investigation. Much is still unknown, but women are no longer satisfied by many of the answers they receive regarding the safety of the factories, offices, laboratories, hospitals, schools and shops in which they and their families work. They want more complete information and action taken to clean up their workplaces, if it is necessary.

The special importance of reducing occupational risks for the pregnant worker is part of the much larger social objective of protecting all women's occupational health and safety. This book was written to help women attain this goal. It aids working women in understanding and making decisions about pregnancy and job-related health problems. It helps them judge scientific data, assess risks, and tells them what kind of information is needed to influence their decisions. It gives them a standard for judging their own work situation, shows how they might improve it and, armed with increased knowledge, seek to improve working conditions for all pregnant women. While this book alerts women to possible dangers in their work place, however, *it is not a substitute for obtaining individual healthcare advice from a well-qualified professional.* Every worker's physiological and genetic makeup differs as does her interaction with the work environment.

Of equal importance is that workers realise that they have to be aware of tomorrow's hazards as well as today's. New technologies can present potential reproductive hazards in the workplace, and these need to be continually monitored and evaluated. Working women – and men – need to understand and use scientific information to promote and preserve both their own health and that of their unborn children. The ability to protect their health during their reproductive years may be as important to them as learning specific job skills.

Women cannot be held responsible for all of the damaging influence of working conditions on their children. Men are exposed to occupational reproductive hazards as well. More and more evidence suggests that men's reproductive organs and sperm are suffering damage in the workplace and that their offspring are suffering the consequences. The act of conception is a joint enterprise between women and men. Freeing the workplace from reproductive hazards should also be a joint venture and men are increasingly becoming involved. Potential fathers will find the information in this book beneficial.

This book draws on information from public health and scientific literature, unions, government documents and public interest groups. But it is the voices of the pregnant women themselves who were interviewed that speak most dramatically of the urgent need to achieve a pregnancy friendly workplace.

Readers have diverse agendas. Each of you may be concerned about different aspects of reproductive hazards in addition to learning about the specific occupational risk you face. You may be distressed about the dearth of scientific information, or you may want to learn about the protection government regulations offer or the success of social action. Therefore, several facets of the problems are covered and each chapter is self-contained. You may just want to read the sections pertaining to your workplace or individual concerns instead of reading the entire book. You may also want to return to other chapters later when you feel the need to investigate further.

Chapter 1 discusses the plight of the pregnant worker and working mother of young children in Great Britain today. Working women may still have difficulty in achieving a safe reproductive health environment and job security despite protective legislative efforts by the British government and the European Community. Even though great strides have been made by women in obtaining management and professional positions, most jobs held by women are still low paying and non-unionised. These women are most vulnerable to reproductive harm and often are not aware of their rights or are afraid of losing their jobs if they assert them. Childcare facilities are woefully inadequate. Women with better paying jobs also face severe childcare problems and require more opportunities for flexible work schedules and career breaks schemes.

Chapter 2, reviewing the biology of reproduction, examines evidence showing that males as well as females face reproductive risks at work which can damage their reproductive systems. These hazards also increase the rate of birth defects and cancer among their offspring and the likelihood of their female partners having miscarriages. The appendix in this chapter also provides information about prenatal diagnostic techniques and genetic testing in the workplace.

Chapter 3 is the heart of the book and addresses the impact of both the physical and social work environment on the pregnant woman. For example, a pregnant woman's body undergoes physical changes which makes her more sensitive to noise and heat. In the final months of pregnancy, her balance shifts and she may be uncomfortable if she has to work for many hours in one position. The treatment she receives from her supervisors and co-workers also influences her well-being. Minor, relatively inexpensive changes can often eliminate stress and discomfort and make work remain a pleasurable and healthy activity.

Women need to know whether their specific jobs pose hazards to themselves or their unborn children. The chapter analyses a variety of work sites – home, office, school, hospital, factory and the service industry, and discusses toxic substances found in some workplaces. This chapter identifies and suggests ways of minimising or avoiding these hazards. The lower paid jobs are emphasised because the health and safety issues are usually more severe in these sectors.

Scientists and consumers view risks differently and managements may try to downplay the hazards involved. This makes it very difficult for workers to judge risks intelligently. Chapter 4 presents information about bacterial, animal and human studies and talks about how to weigh the risks and benefits. It focuses on ways to close the information gap. Obtaining information and being confident of its accuracy is a major problem for the pregnant worker. The media often exaggerate stories while corporations frequently hide information. Even when adequate data are available, the individual may not know how to put information together to obtain a clear picture of the situation. Once pregnant women familiarise themselves with scientific terms and learn how risks are analysed, they can understand the dangers they face at work and can question and counter explanations, denials and claims made by their employers.

Several initiatives taken in the United States and other countries provide useful insights and valuable suggestions about ways of making work safe and healthy. Chapter 5, the final chapter, recounts some innovative approaches taken by women, as well as successful organising and lobbying techniques. It also provides information about how to achieve action from official bodies, including your local authority.

The glossary, appendices, suggested occupational health references (some of which are easily understood by a lay person while others would require a scientific background) and addresses of organisational resources at the back of the book furnish definitions of commonly used terms in the occupational health field and provide up-to-date sources of information. This includes, in Appendix A, an occupational information form for you to fill out and give to your doctor – only after women are armed with knowledge about the reproductive health risks from their occupational exposures can they then decide what steps they may want to take to avoid them.

This book can be referred to repeatedly in the journey towards achieving a pregnancy friendly workplace.

ACKNOWLEDGEMENTS

I owe particular thanks to Jenny McLeish and Jan Fry of the Maternity Alliance whose work in adding and updating the legislation regarding health, safety and pregnancy in the UK, co-writing the appendices and checking for accuracy made this book possible. As a 'friend from across the ocean' this was very difficult for me to do alone. I also want to thank my colleagues, friends, students, family and professionals in the fields of occupational and women's health in both the United Kingdom and the United States whose suggestions and support made this book possible. I am especially grateful to Dr Caroline Berry, Division of Medical and Molecular Genetics at Guys Hospital who obtained permission for me to interview pregnant working women. Claire-Marie Fortin and Sue Barlow deserve particular thanks for their support of the orginal idea for this book. Daphne May provided invaluable help in collecting information pertaining to workplace issues in the UK and co-authored sections of Chapter 5 ('Action Through Official Bodies' and 'Action Through Your Local Authority'). I am also grateful to the Faculty Institutional Leave Committee of the College of New Jersey which rewarded me a sabbatical and additional released time from teaching to work on this book.

1

The Right Mix: A Working Woman's Agenda

The Factory Girl (Stanza 4)
Ten hours a day of labor,
In a close, ill-lighted room.
Machinery's buzz for music,
Waste gas for sweet perfume;
Hot, stifling vapors in summer,
Chill drafts on a winter's day,
No pause for rest or pleasure
On pain of being sent away;
So ran her civilized serfdom –
FOUR CENTS an hour the pay!

J.A. Phillips, *Machinists' Monthly Journal*, September 1895, p.3.

One hundred years later much has changed, but parallels remain. Workplaces can be excessively noisy and the temperature is often too hot or too cold. Many workers still breathe unhealthy air, but now it is called indoor air pollution or tight building syndrome. Women are still pushed to work faster, but now supervisors often use computers to check on workers' accuracy and speed.

While wages of four cents an hour have disappeared long ago, women's wages in Britain today are still low compared to those that men receive. According to the Equal Opportunities Commission (EOC), pay differentials between men and women are greater in the United Kingdom than in most other European Union (EU) countries and ironically, this is still the case even though more women work in Britain than any other EU countries except Denmark.

WOMEN: MAIN CONTRIBUTORS TO FAMILY INCOME

As was the case with the factory girl, most women work out of economic necessity, either to increase the low wage of a partner or as sole breadwinner. A government working paper pointed out that the number of families in poverty would treble if it were not for the wife's earnings. In several hundred thousand families, the male partner is unemployed so the women are the sole earners. Moreover, one in eight households is now headed by a single parent, and 91 per cent of these are women. In fact, 90 per cent of the overall increase of 3.1 million in the civilian labour force between 1971 and 1989 was attributable to an increase in the number of women. By the year 2000, women are expected to make up 45 per cent of the total civilian labour force.

Between 1984 and 1994 the percentage of women with a dependent child (or children) who were employed increased from 55 per cent to 64 per cent and the likelihood of such women being employed increased as their youngest child got older. The most marked difference between 1984 and 1994 was for women with very young children (under 5 years of age). Thirty-seven per cent of these women were employed in 1984. This rose to 52 per cent in 1994.

While these changes are dramatic, research carried out by the Institute for Manpower Studies suggests that if good and affordable childcare was available, about another million women would also enter the labour market.

> In this country, basically women are discouraged from returning to work after they have a baby. Work arrangements, nursery facilities – hardly an employer has them. Hours of work discourage women from returning to work.
>
> Sonia, self-employed textile worker

Much of the part-time work available to women in the UK is portrayed as fitting in with women's domestic commitments. But more often it disrupts family life and is primarily suited to the requirements of employers. One example of this is Sunday trading. While employers claim that Sunday work is voluntary, there have been a few cases where the workers, mainly women, were told that they had to work some Sundays if they wanted to keep their jobs. Violations were eventually addressed by enactment of the Sunday Trading Act which came into force in August 1994. The Act (now part of the Employment Rights Act 1996) sets out the protection concerning Sunday working which is enforceable through industrial tribunals. This means that shop workers should not be dismissed, made redundant or subjected to any other penalties if they choose not to work on Sunday.

Another way employers have saved on overheads and wages is by 'putting work out', by which women then do the work in their own homes. This sounds much better than it is. These women are not eligible for many benefits, get paid less and have irregular schedules. They can go from working long hours without time off in order to meet very short deadlines to going without any work for extended periods. Rather than being able to pace their work in order to meet their own family's needs, the women who work at home frequently have less flexibility than women working outside the home.

THE PREGNANT WORKER

Not too long ago women were forced to leave their jobs as soon as their pregnancies became evident. These discriminatory employment policies were based on myths rather than on a reliable body of scientific information. Men were embarrassed by pregnant women's bodies and women were embarrassed to be seen publicly during their last months of pregnancy. Even though medical opinion now believes that a healthy pregnant woman can perform her normal work tasks throughout most, or all, of her pregnancy, some males (and females) still would rather not see pregnant women in the workforce for reasons of propriety or for the adjustments that might have to be made.

> Male managers should have training sessions about pregnant women in the workplace. Their attitude was that they only saw women in the workplace as workers even though they have wives and children. There should be more breaks for women to put their feet up. When they put on teas and coffee during breaks, there should be herbal tea, milk and orange juice.
>
> Amira, librarian

At present, most British women ignore the remaining discrimination against them and work through part or all of their pregnancies. A little over half the pregnant women in the UK and 74 per cent of women with first pregnancies are in the workforce.

Until 1975 it was legal to dismiss a woman for being pregnant. The 1975 Employment Protection Act was supposed to protect women against unfair dismissal because of pregnancy, but it was hard to prove that pregnancy was the cause, and many women were not covered by the Act. Since October 1994, under the new laws implementing the European Union Pregnant Workers Directive, it has automatically been unfair dismissal to sack *any* woman because of her pregnancy and, in addition, an employer who sacks a woman because she is pregnant is also breaking the sex discrimination laws. However, while these laws make it more difficult for a pregnant worker to lose her job, a wily employer can still get around the rules, especially when there is inadequate enforcement.

Despite the persistence of incidents of discrimination, working conditions and provisions for pregnant staff continue to improve slowly. New Ways to Work, an educational charity which promotes flexible work arrangements (details are given in 'Finding Out About Different Options' later in this chapter), cites a survey showing that in 1992 nearly 62 per cent of 175 companies questioned stated that they were prepared to alter working arrangements to meet their pregnant employees' needs – a 15 per cent increase from 1990. There was also an increase in the number of companies offering more than the statutory maternity pay. This is the good news. The bad news is that the increase was only 2 per cent. Seventy-seven per cent of the companies surveyed still only offered minimum statutory pay and leave.

WORRYING ABOUT REPRODUCTIVE HAZARDS IN THE WORKPLACE

While all workers need to adjust to the requirements of the employer, the pregnant worker may have a more difficult time and will worry that some aspect of her work may injure her foetus. Almost every woman planning to have a baby asks herself whether her baby will be all right. It is different today, however, from the time when poor women had no protection against reproductive hazards in the workplace while working middle-class, educated women either did not have children, left work when they became pregnant or believed, sometimes erroneously, that they worked in a safe environment.

Furthermore, women are no longer resigned to working under conditions that might cause their offspring to be stillborn, to not survive to adulthood or to be born with a severe physical or mental disability. Instead, couples today expect to have a small number of 'quality' children – perhaps an unrealistic expectation whose fulfilment is too frequently thwarted. The media has given a great deal of attention to the number of newborns being born with birth defects, to the rise of infertility and to heartbreaking stories of attempts by infertile couples to use new reproductive technology in order to have a 'child of their own'.

The causes of many of these reproductive problems are still unknown and widely debated. But one of the most likely culprits responsible for miscarriages and birth defects is thought to be the exposure of the mother during pregnancy (and, in some instances, the father before conception) to one or several toxic substances or agents. While some of these unfortunate outcomes are attributed to infections or medications, others may be due to workplace and environmental hazards.

The worker contemplating pregnancy must resolve to her own satisfaction 'the pregnancy connection' – the link between her work and reproductive harm. She faces three tasks:

- to try to protect herself from occupational reproductive hazards
- to secure equitable pregnancy treatment and adequate maternity benefit
- to obtain good childcare arrangements if she wants to return to work.

Sometimes very simple changes will suffice to protect the health of a pregnant worker. At other times, more formal guidelines are called for.

> I empty bins, tidy classrooms, pick up anything dropped, and wipe blackboards. There was some bending, climbing up and down stairs and moving about, but it was not strenuous. There was not a vast amount to do. I could take my time and sit down if I felt like it. I moved chairs and tables before I was pregnant. When I knew I was pregnant, I refused to do that. They have plenty of men to do that.
>
> Emma, part-time cleaner

At the moment the rules and regulations regarding maternity are being written. We changed from the NHS to a Trust last year. I haven't actually seen the written document, which would be helpful. I was talked through it.

Heather, hospital clinic administrator

Before October 1994 the law provided employed pregnant women and mothers who just had babies with certain rights, but there was a catch. There were exceptions depending on length of employment and earnings. Not all working women had these rights and many who most needed the protection were not covered by the legislation. For others, the coverage they were entitled to was woefully inadequate.

THE EUROPEAN UNION PREGNANT WORKERS DIRECTIVE

Table 1.1 Maternity leave provisions in the European Union prior to the adoption of the EU Pregnant Workers Directive

Country	*Maternity Leave*	*Pay*
Belgium	14 weeks (6 before birth)	100% for 1–4 weeks, thereafter 80%
Denmark	28 weeks (4 before birth)	90% of salary
FRG (Germany)	14 weeks (6 before birth)	100% of salary or fixed sum
Greece	15 weeks (6 before birth)	100% of salary
Spain	16 weeks	75% of salary
France	16 weeks (6 before birth)	84% of salary
Ireland	14 weeks (4 before birth)	70% of salary
Italy	20 weeks (8 before birth)	80% of salary
Luxembourg	16 weeks (8 before birth)	100% of salary
Netherlands	12 weeks (6 before birth)	100% of salary
Portugal	90 days (6 weeks before birth)	100% of salary
UK	40 weeks	6 weeks at 90% of salary and 12 weeks at a fixed reduced sum.

Source: Protection at Work of Pregnant Women and Women Who Have Recently Given Birth – A Proposal for a Directive from the European Commission, Annex 1 Maternity Leave.

In October 1992, the European Union (EU) adopted a Pregnant Workers Directive (see Appendix E at the back of this book) and in October 1994 new maternity rights provisions implementing this Directive came into force in the UK. These rights are now contained in the Employment Rights Act 1996 and the Management of Health and Safety at Work Regulations 1992 .

So, are you much better off under the new UK provisions than you were before? Does the new law apply to you? What does it cover? The basic provision is set out in the boxes below but more details about your rights under the new legislation are given at the end of the book: Appendix C for health and safety and Appendix F for other rights. For the latest information send a stamped, self-addressed envelope plus a cheque or postal order for £1.00 to the Maternity Alliance, 45 Beech Street, London EC2P 2LX and ask for the leaflet *Pregnant at Work*, or ask at a Job Centre for the Department of Trade and Industry booklet *Maternity Rights*, reference PL958. These two pamphlets are written for the pregnant worker, but the Health and Safety Executive has also written one for employers: *New and Expectant Mothers at Work – A Guide for Employers*, available from HSE Books for £6.25 (PO Box 1999, Sudbury, Suffolk CO10 6FS. Tel: 01787 881165).

The new law applies to you if you are pregnant, breastfeeding or have given birth in the last six months. Most of the rights only apply if you are an *employee* (not self employed).

Maternity leave: All employed pregnant women who give the right notice to their employer are entitled to at least 14 weeks of maternity leave no matter how long or for how many hours a day they have worked. Two weeks of these are compulsory and must be taken immediately after giving birth. You have the right to your own job back afterwards. The old rules about the 'right to return' also remain in force, so if you have at least two years service with your employer by the end of the twelfth week before the week the baby is due, you can take up to 29 weeks off from the week the baby is born and have the right to return to the same or a very similar job.

The complicated notification rules for maternity leave mean that some of you who want to return to work lose your right to do so because of the confusing red tape involved.

Maternity pay: Provided that you have six months service by the time you are about six months pregnant, and provided you earn enough to pay National Insurance, you get Statutory Maternity Pay. This is 90 per cent of your average wage for six weeks and then a flat rate for up to twelve weeks.

An estimated 380,000 British women a year are in paid employment while pregnant. Of these, according to the Equal Opportunities Commission, about 40 per cent did not qualify for Statutory Maternity Pay under the old regulations because they earned too little, did not have sufficient service or worked part time. In 1995 the Employment Protection (Part-Time Employees) Regulations abolished maternity pay discrimination against part-time workers, so pregnant women who work less than 16 hours a week now have the same rights as those who work for more than 16 hours a week. But the estimated two million low-paid women who do not earn enough to pay National Insurance are still currently excluded from the maternity pay system and receive nothing. The Maternity Alliance is mounting a legal challenge to this exclusion.

The UK blocked the more generous pay and leave provisions which were originally proposed in the Pregnant Workers Directive. In most EU countries, maternity pay is at least 80 per cent of wages.

Antenatal care: As was already the case, you are entitled to paid time off from work for antenatal appointments, including parentcraft and relaxation classes.

Dismissal: An employer cannot dismiss you for any reason connected to your pregnancy or maternity leave. If you are dismissed during pregnancy or maternity leave your employer must give you written reasons for the dismissal.

Health and safety: There is also a health provision. Employers must assess their workplaces to see whether there are any risks to your health and safety or that of your baby and if a significant risk is found they have to inform you and decide what measures should be taken. Hazards include mental stress as well as biological, chemical and physical substances and agents. The employer must then take certain steps:

- do all that is reasonable to remove or reduce the risk
- if the risk remains temporarily alter your working conditions and/or hours, if this is reasonable and would avoid the risk

- if this is not reasonable or would not avoid the risk, offer you a suitable alternative job
- if this is not possible, suspend you from work (granted paid leave) for the period needed to protect you and/or your baby. All your contractual employment rights continue during the period of leave.

Night Work: Pregnant women working at night also benefit under the new law. If you do night work and have a medical certificate from your doctor or midwife saying that this is detrimental to your health or safety, your employer must offer you suitable daytime work or, if that is not reasonable, suspend you from work (grant you paid leave).

MAKING THE LAW WORK FOR YOU

Laws offer protection, but only if they are enforced. Sometimes working women, particularly pregnant workers, feel too vulnerable to ensure that their supervisors comply with the legislation. For example, every pregnant worker in the UK is entitled to paid time off to go to the antenatal clinic; all you have to do is request time off to keep an appointment. After the first appointment, you need to obtain a medical certificate from your antenatal clinic confirming that you are pregnant. Most employers obey the law, but a few refuse to abide by it. Usually this happens when the woman is fearful of losing her job if she protests.

> Most of my patients do not have a problem in getting time off to come to the antenatal clinic, but a few of them cannot get off from work. I fit these into my evening surgeries. When an employer insists that a woman performs work that I believe to be harmful to her pregnancy, I write her a sick leave certificate. Most male doctors are not so sympathetic.
>
> Dr Mary Black, general practitioner offering shared pregnancy care

> A number of pregnant women have problems with their employers. If there is trouble with an employer and he won't let them go for their antenatal care appointments during work hours or refuses to

remove them from hazardous work, the unit puts the women in touch with the Citizens Advice Bureau. There is not much else they can do.

Geraldine, nursing sister in a hospital antenatal clinic

If your employer refuses to pay you for the time off for antenatal appointments, you can bring a complaint to an industrial tribunal. These tribunals are informal courts which deal with employment disputes. They are designed to be accessible for individual workers but bringing a case can be confusing and time consuming. Unless you have prior experience in tangling with officials, you may need an advocate to help you through the legal maze. Ask for help from your safety representative, union, the local Citizens Advice Bureau, law centre or some of the specialised support groups, charities and organisations listed at the end of the book.

You can also bring a case to an industrial tribunal if you are sacked or unfairly treated for being pregnant or for any reason connected with your pregnancy, or if your employer refuses to let you come back to work after maternity leave. For enforcement of your health and safety rights see Appendix D.

Many regulations provide some leeway and a few employers take advantage. The majority of employers abided by the old law, but some twisted the meaning of 'reasonably', 'practical' and 'suitable' when it came to reinstating women after maternity leave. They were prejudiced against hiring women with young children. Only time will tell whether these die-hards will still try to twist the new laws to their liking.

Some of these firms that try, illegally but subtly, to discriminate against pregnant workers and young mothers seem to divide women into two categories – those who are mothers and those who are not. Even women who were highly thought of and treated well when they did not have children are at times treated shabbily when they became mothers. One father described what happened to his daughter-in-law who worked in the financial sector: she had received several promotions, more than some of the men, but when she tried to return to the company after maternity leave she was offered an inferior job which she turned down. The firm did not present her with another offer. She was told that as she turned down their first job offer, they felt no further responsibility.

As we can see, what the law says should exist and what really does exist is not always the same thing. There is usually a lag between a law that changes a long-held position and employers and supervisors accepting the changes. A survey carried out by the Regional Research Partnership in 1989 for *New Woman* magazine found that 24 per cent of employed women said their promotion prospects had worsened as a result of becoming a mother; 12 per cent were given the impression that they would be better off if they left their jobs and became full-time mothers; 17 per cent of their male bosses made snide and sarcastic remarks about pregnant workers and working mothers; and 15 per cent feared that they would be replaced permanently while on maternity leave. The Midlands had the highest percentage of employers who disapproved of working mothers and who were particularly unsympathetic to those that did return to their jobs.

MATERNITY PAY

Women vary in their attitudes toward the amount of maternity pay for which they are eligible. Some women, particularly in lower paying jobs, are glad to get the statutory minimum. Others are ambivalent about the low level of maternity benefits they receive even if they are above the minimum.

> I get the local authority maternity benefit – six weeks full pay, twelve weeks half pay and the rest until twenty-nine weeks without pay. It is not really adequate, but I'm lucky to get them and I'm happy to have them.
>
> Amira, librarian

> I've worked for the firm for four years and they are only giving me statutory benefits. They give a better package to managers – those working for five years. I am pushing for a better package. I am the main breadwinner and I cannot afford to stay on statutory. Raising the benefits would mean nothing to them. It is a multimillion dollar firm. Only recently have women been chartered accountants. In a department of 300 accountants, I think I'm only the third woman to become pregnant.
>
> Catherine, chartered accountant

The question of maternity pay, at least for the immediate future, has been settled by the legislation implementing the EU Pregnant Workers Directive, but other benefits such as parental and dependency leave, childcare and non-traditional work schedules and careers are still woefully inadequate. Maternity pay is a major issue for working women, but it only covers a short period of time; combining a career and family, simultaneously or sequentially, means juggling obligations and responsibilities for years. Non-traditional work schedules would help working mothers enormously.

Ninety-three per cent of the women in a 1991 Institute of Manpower Studies (IMS) survey felt that time off for domestic emergencies, particularly when regular childcare provisions broke down, was the most helpful form of flexibility.

CAREER BREAKS AND NON-TRADITIONAL WORK SCHEDULES

Career Breaks

Some companies offer long career breaks which can be taken for any reason, such as taking care of children, elderly or disabled relatives or going back to college for further study. The term 'career break' usually refers to a formal arrangement for extended leave. It involves an understanding on both sides that the employee will return to the company at the end of the agreed time to the same or equivalent job. Generally, career breaks are unpaid and can last up to five years, though sometimes they only last several weeks or months. While they are available to both male and female workers, it is the female workers who most often take advantage of this option.

Most of these schemes have some built-in training or short periods of paid work. Sometimes career break staff are used for two to four weeks a year to cover sickness or holiday leave taken by other employees, which has advantages for both employers and workers. The workers keep in touch with new developments at their companies and so do not become out of date, while companies benefit by gaining a knowledgeable replacement. Barclays Bank, GEC-Marconi, Shell and several other major companies have career break schemes.

Job Sharing

Job sharing consists of one full-time job held by two people each working part time. They work either several hours a day or certain days a week. With a little planning and coordination this option works well. The employer receives reliable service; the job sharers hold down a more interesting job than the usual part-time one, have fewer problems with childcare, and can spend more time with their children. Each is eligible for a proportion of the benefits accruing to the full-time job.

More than half the job sharers are mothers or fathers of young children. Commonly, job sharers already know each other and often team up after returning from maternity leave around the same time. A person looking to share a job also may place an advert in a local newspaper or professional or trade journal. Some large employers run their own job sharing registers and New Ways to Work administers a job sharing scheme in the London area.

A commonly asked question about job sharing is what happens when one of the job sharers wants to leave? Some employers revert the job to full-time status and others simply advertise a job share vacancy. The pool of potential job share recruits is high and there should be little trouble finding someone who is satisfactory to both the existing job share partner and the employer.

More employers are beginning to accept job sharing, though up to now it was often dismissed as being impractical. Employers felt that it might be feasible for other companies, but not for their own. Job sharers now work in shops, schools, banks, factories, laboratories, local councils, the BBC, the Stock Exchange, the House of Commons, the Home Office and the NHS. According to their experience, most jobs can be shared. For example, Leicester City Council instituted an imaginative scheme in 1986: every council job is regarded as a potential job share, unless specific reasons for its exclusion have been approved. Very few jobs have been accepted for exclusion.

Job sharing benefits the employer as well as the worker. It retains skilled personnel, reduces staff turnover and each shared job benefits from the expertise and experience of two people.

> Childcare arrangements will probably be a problem. There is no childcare at work, no crèche, no part-time work, and no job sharing. I know, because a few of us asked. A couple of us tried to get job sharing, but they said no. If I can't cope with going back full time, I may try to get another job where I can job share, but a lot of companies won't take you on if you hadn't been there.
>
> Diana, secretary at a large investment bank

Obviously, the disadvantage is the loss of income which is necessary for many two-income families as well as for women who are the main breadwinners.

Flexitime

Flexitime should not be confused with working the swing shift, night shift or weekends out of necessity. Many women reluctantly work these hours either because they need the higher pay, their employers require it, or they can not afford childcare costs and their husbands are home to take care of the children at these times. Flexitime and the compressed work-week are voluntary choices and, with goodwill on both sides, pregnant workers and employers could work out many more kinds of arrangements. Pregnant workers and working mothers greatly value the ability to change start and finish times of their working day. This creates greater job satisfaction and results in stress reduction. As stress is considered to be a reproductive health hazard, ways to eliminate it bring additional health benefits.

> I work flexitime and have to average seven hours and twelve minutes per day for the week. I can start between 7.30 a.m. and 10 a.m. and quit between 3 p.m. and 7 p.m. Lunchtime is between 11.45 and 2 p.m. I have a flexisheet and I keep record of my time. My manager checks it every once in a while. I can build up a flexiday and I'm allowed to be eleven hours in debit per month without anyone saying anything about it. It gets carried over to the next month. I fit my breaks in when I want them. I have my own office and I can shut the door and rest for a few minutes. Some days I might work a long day and another day work for half a day.
>
> Zarina, clerical assistant

> I worked as a catering supervisor in a university during my pregnancy. I had high blood pressure during pregnancy. I came to work at six in the morning and when it got too hot, I took my paperwork home and did not come in for the rest of the day. If I had not had that choice, I would have had to take time off from work.
>
> Maria, catering supervisor

Part-time Work

Part-time work is an important option for women raising a family. There is an urgent need to raise the status and pay of part-time workers and to extend it to more senior posts. Sometimes, however, people who formerly worked full time and return to a three- or four-day schedule find that they are expected to do five days work in the new shorter schedule.

> They do keep your job for you and you can make arrangements to work part time. I was thinking of working four days a week when I return, but then I thought that I would have to work late every Thursday night to finish the work that would have to be done on Friday.
>
> Catherine, chartered accountant

Many women want the option to return to their old job after maternity leave, but on a part-time basis. Although there is no statutory right to do this, recently women have won industrial tribunal cases after their employer has refused to let them work part time or job share, as it has been seen as a form of indirect sex discrimination (as more women than men have childcare responsibilities which prevent them from working full time). The employer has to justify objectively why the job cannot be done on a part-time or job share basis. The law on this is complicated so you should always get advice: contact New Ways to Work or Maternity Alliance (addresses given later in this section and in the Useful Addresses list at the end of the book).

Annual Hours Contracts

Some companies and local authorities are trying out annual hours contracts in which the employees and employers agree to work longer or shorter working days or weeks or a specified number of days each month as long as the total number of hours agreed to in the contract are worked each year. Both workers and employers can benefit. These contracts give individual employees the flexibility to attend to family needs and ensure the employer that the work gets done. The employer obtains a more efficient labour force as the employees are not spending their work time worrying about their difficulties at home. Under an annual hours contract, employees would seldom have to take days off from work with the possible loss of pay in order to take care of family problems. They also gain peace of mind knowing that the structure of their everyday lives allows them to meet both their personal and workplace obligations.

Finding Out About Different Options

Some UK firms, local authorities and hospitals have considerable experience about flexible work arrangements which could be translated into a national policy for the country. Sometimes, as the old saying goes, 'necessity is the mother of invention'. Several hospitals suffering from severe nursing shortages decided to experiment with job sharing and other types of alternative schedules. Some of the self-governing NHS hospitals have undergone drastic employment policy changes. The Bradford Trust, for example, in return for greater flexibility among its workforce, was willing to consider such innovations as childcare vouchers, paternity leave and adoption leave.

The educational charity New Ways to Work (309 Upper Street, London N1 2TY, Tel: 0171 226 4026) promotes flexible ways of working. They provide information, publish fact sheets and booklets, run training sessions and seminars, operate a consultancy service and carry out research. Some of the options they present are: flexible working hours, part-time work, job sharing, term-time working, career break schemes, voluntary reduced work time options, sabbaticals and working from home (telecommuting).

Flexitime – varying arrival and departure time – and the compressed work-week where longer hours are worked on fewer days are the two most common plans.

While it will probably be a long time, if ever, before any of these 'family friendly' working practices becomes a standard option in the UK, at least they are now being talked about as realistic options and being tried at more forward looking workplaces. According to the 1993 Labour Force Survey:

- 12 per cent of employees (2.6 million) worked flexitime – 10 per cent men and 14 per cent women
- 9 per cent of employees (2 million) worked a system of annualised hours (most common among professionals, particularly teachers)

- 5 per cent of employees (1.1 million) worked term times only, but only 1 per cent (0.27 million) of these worked outside of the education sector
- 4 per cent of full-time employees (0.7 million) worked a compressed working week.
- 4 per cent of part-time employees (0.19 million) job share.

What is interesting about these statistics is that the proportion of women as part of the flexible workforce has remained fairly stable at about 50 per cent since the early 1980s, whereas the past decade has seen a 9 per cent increase in men's participation which, in 1993, reached 27 per cent of the flexible workforce.

CHILDCARE WOES

Before we even give birth to our babies, we begin to think about problems we face when we return to work and question whether we should return at all. Regardless of the reasons for working – self-fulfilment, economic necessity, or both – being separated from our newborns is a wrench and obtaining childcare a worry.

Many women want the option of staying home with their baby for the first year or two, or at least for several months, without losing their jobs or seniority. The UK, however, has no parental leave legislation pertaining to care for small children or dependency (family) leave legislation dealing with care for sick children. There is only the extended maternity absence of up to 29 weeks from the week the baby is born, available to women with more than two years service. Parental leave of between six months and three years exists in most other European Union countries. In 1996 a European Union Directive on parental and family leave was adopted under the Social Chapter, requiring that by 1998 all countries except the United Kingdom must introduce a minimum of three months parental leave for both men and women and some period of dependency leave. The UK was exempt as it had opted out of the Social Chapter but, in May 1997, the new Labour government opted back in.

Staying at home to look after children is still not an option for the many women who need their incomes in order to help support their families or who are the sole breadwinners. In 1987, a Centre for Economic Policy Research discussion paper, *The Cash Opportunity*

Costs of Childbearing, found that it cost a woman between £120,000 and £135,000 in lost earnings (an appallingly high sum) if she had to stop working because of childcare problems and then had to return to lower-level work.

Parents who have to or choose to work question the advisability of leaving their young children in someone else's care while they go to work and wonder what kind of care is the best. The Thomas Coram Research Unit set out to answer this question. The unit is following children from their infancy who are in different kinds of care – with mothers, relatives, childminders and nurseries. Some evidence indicates that the wide social network and variety of experiences provided by a nursery would be beneficial for children. But because the study follows children for several years, the answers will not be known until the late 1990s.

Finding good quality, affordable childcare is a real problem for many families. Figures compiled by the Daycare Trust show that there is only one registered childcare place for every nine children under eight years of age. Several factors exacerbate the situation: the majority of childcarers receive very low pay, training opportunities are inadequate and chances for career advancement limited. A childminder caring for two children for 40 hours a week earns less than half the average pay for non-manual women workers.

Going back to work shortly after giving birth and continuing to work while raising young children is a constant juggling act and in times of crisis, one of the balls may be dropped. Mei, a recent immigrant from Vietnam, is a good example of such juggling. She was expecting her mother-in-law to come and take care of the baby for a while, and planned to return to work in the small London restaurant she and her husband owned four weeks after giving birth as she could not afford to stay out of work any longer. While Mei and her husband had not planned all their childcare arrangements, she was also counting on her sister- and brother-in-law, who lived nearby, to babysit part of the time, and planned to take the baby to work if necessary. A nanny was too expensive, but she thought she could hire some babysitters in the evening when she worked the most.

Both workers and employers find that conflicts between jobs and pressures from family duties are becoming increasingly troublesome. At first, because the conflicts were perceived only as women's problems, they were largely swept under the rug. Now, these conflicts are reaching the surface and being openly discussed, perhaps because they are affecting male workers as well. Most family men have working wives with their own employment obligations; mothers can no longer be counted on to stay home with a sick child if both incomes are needed. Fathers then have to stay home, arrive at work late, or leave early to fulfil their family obligations. Often mothers and fathers have to alternate taking days off when children are sick so that neither loses his or her job.

> If my child is sick, my husband and I each take days off from work or my in-laws watch him. He goes to a childminder after school.
>
> Maria, cafeteria supervisor

These strains have led to high absenteeism and losses in productivity, forcing the business community to emerge from its cocoon and search for solutions.

TYPES OF CHILDCARE AVAILABLE

You can piece together many kinds of childcare arrangements – nannies, relatives, friends, neighbours, childminders in your home or theirs as well as crèches, nurseries and out of school childcare arrangements – but not all of these are always available, satisfactory or affordable. The lack of childcare schemes is a problem of enormous importance in the UK as it provides fewer publicly funded childcare centres than its European partners. (To find out about the childcare options in your local area, contact your local council.)

Yet the lack of formal childcare is not the only problem. A recent survey found that the demand for fathers to help look after the children far outstrips the supply. Fathers had few childcare responsibilities and a majority of those with children under twelve agreed with the statement that the husband should be the breadwinner and the mother should look after the children. Needless to say, only one-quarter of the mothers agreed. Even if more of the mothers wished to stay home and look after the children, they could not do so as they need the money to help support their families.

But if you are a single mother or a mother with a partner who does not want to share childcaring responsibilities on a more equal basis, you need good and affordable childcare *now*! At least childcare has reached the top of the national agenda, and several groups – such as

Parents at Work, The Daycare Trust and Working for Childcare Consultancy Service – as well as businesses and local councils have grappled with the problem and have come up with several different types of schemes (addresses are given at the end of the book).

Local authorities have tried to meet some of the childcare needs, but are severely limited in what they have been able to provide. The nurseries they run are usually priced reasonably compared to private facilities, but they have places for fewer than 2 per cent of all children under two. Unless a child is defined as being 'at risk' by the local social services department, it is almost impossible to obtain a place in these nurseries. Occasionally a single parent might be lucky.

The Childcare Circle

One innovative idea to solve the joint problems of shortage of childminders, childcare facilities and the cost of childcare is called the 'Childcare Circle'. Mothers work on rota four days a week but are paid for five days. On the fifth day, each in turn looks after her own and other members of the Circle's children. There can be problems with this however: having a different childminder each day of the week is a disadvantage even though continuity of care does occur over a longer period and, for the women concerned, caring for five children on their day off from their regular jobs may be more stressful than working the normal five day week.

Joint Ventures in Childcare

There are several ways of reducing the cost of childcare for companies who are willing to venture into this new area of company-sponsored childcare. Partnerships are one of the main ways. Commercial childcare providers will open new nurseries if they are guaranteed that their openings will be filled by children of workers from one or more companies. The NHS has opened several partnership nurseries sharing facilities with Midland Bank employees, for example. Several smaller companies in one geographical area can also jointly operate one facility for their employees, sharing space, staff and expenses. Women can then visit their babies during their lunch hour and continue nursing if they wish.

Some local authorities, such as Leicester City Council, are establishing nursery partnership schemes with interested employers. Other government agencies are also getting involved.

Cash Allowances and Childcare Vouchers

Many companies in Britain are attempting to meet the childcare needs of their employees by giving cash allowances or childcare vouchers. In order to be eligible to be paid with a childcare voucher, the childminder must either be registered with the local authority or be someone who is exempt from registration requirements, such as the child's grandmother. An example is the Burton Group, which contributes to the cost of their staff's childcare through childcare vouchers.

Single Employer Nurseries

If there is enough demand from employees in one large company, a commercial childcare provider may find it possible to set up a nursery for the exclusive use of the employees of that company. The company is saved the capital outlay and is free from the responsibility of managing the nursery. American Express is one such company that took this route. It guaranteed places for 50 children, many of whom had to be looked after from 8 a.m. to 6 p.m. American Express subsidises the facility and parents pay according to their salary grade.

WHAT CAN BE DONE?

Childcare remains a severe problem for most working mothers. One study, published by the Institute for Public Policy Research, proposes several alternative models for designing a national childcare policy. Their cost-benefit analysis predicts that up to 500,000 children could be raised out of poverty if childcare provisions were adopted. Furthermore, this wouldn't cost the government any money as estimated extra tax revenue generated by investment in these services would range from 5 to 51 per cent. At present, a single mother with no relatives living nearby who could help out would have to ask charities for money for childcare in order to take college courses needed to obtain a decent job at a living wage.

> Childcare is a major problem, mammoth, paramount. I put my name down on the list for the local authority's staff nursery. It cost £74 a week. I don't know whether I can afford that. Where I live the local authority nursery for the public is half the price, but I'm not sure of the standards. I am finding out about childminders as well. I prefer a nursery because he can socialise with other kids.
>
> Amira, librarian

A survey by Working for Childcare found that 75 new workplace nurseries opened in 1991, bringing the UK total to 450. A fourth of these are partnership schemes between the private and public sectors. Despite these gains, only one child in 300 attends a workplace- or employer-sponsored nursery. Working parents who are lucky enough to live in the south east where half the nurseries are located may have a chance of working for an employer who provides such care. Compared with this, Scotland only has 3 per cent and Wales 4 per cent of the total UK workplace connected nurseries. Furthermore, most of the nurseries are in the public sector, sponsored by local authorities, hospitals, colleges and the civil service. Only 16 per cent are in the private sector and 23 per cent are partnership schemes which combine public sector professional expertise and private sector funding.

Childcare Services for the Sick Child

One of the biggest problems that parents employed outside the home face is the breakdown of everyday childcare arrangements. This usually occurs when the child is sick or the childminder has to cancel. Now that two incomes are usually necessary to keep up a desired standard of living, this has become the father's problem as well – the mother cannot be the one that always takes time off from work. To meet this need, in 1991 the BBC World Service started the UK's first in-house emergency childcare facility – taking up to four children aged two and a half to fourteen each day and at no cost to the parent. The emergency service has been used most days, and the World Service believes that it only takes just over one child per day for the plan to pay for itself in terms of staff productivity. While emergency childcare service is a rarity in the UK it is becoming a little more common in the United States, where some programmes have isolation sections in nurseries for ill children, and others send employer-paid childminders to the employees' homes.

New Schemes

Some foresighted companies are even changing their employment policies to meet the needs of women trying to combine family and careers. Tesco is providing a comprehensive flexible working package and is experimenting with alternative schemes. Shell UK has a career break scheme in addition to providing six months' maternity pay. More than 80 per cent of women taking maternity leave from Shell return to work. Likewise, Rank Xerox has increased the numbers of women who return to work after maternity leave from less than 20 per cent to more than 80 per cent by introducing improved maternity benefits and a 'phased return' of part-time working.

The charity Business in the Community was also concerned about the status of women in higher level jobs and as a result, at the end of 1991, it launched Opportunity 2000. The goal was to stem the waste of talented women in business. More than a dozen chairmen and chief executives of large corporations pledged that they would meet specific goals aimed at improving advancement for women at the top. Very few women in the UK hold top executive or management positions in corporations or the civil service or are barristers or judges. In 1991, only 43 out of 688 permanent deputy and under secretaries were women. According to the Association of First Division Civil Servants, the trade union for senior government officials, the number had barely increased in the past ten years.

The lack of good childcare facilities is probably the biggest barrier preventing women from participating fully in all levels of the labour market. The Equal Opportunities Commission has been pushing for more childcare nurseries and argues that providing nursery care is cost-effective compared to the loss of trained staff. The Midland Bank came to the same conclusion – and banks do not tend to be altruistic. After consulting the Institute of Manpower Studies, they decided that the cost of replacing a conscientious employee was equivalent to a year's wages and benefits for the worker being replaced.

Employers who have investigated childcare needs of their employees and ways of providing it have found that they can subsidise childcare in many different ways. The two main paths are by helping their employees with the cost of care the employees find themselves or by the company providing the facilities. The second alternative has the further advantage of providing additional nursery facilities that are desperately needed.

The Need for After-School and Holiday Care

Lack of nursery care and crèches are not the only problems for working mothers. After-school care and holiday play schemes for children aged five and over are also inadequate. It is hard to keep your mind on your job when you are wondering where your school-age children are and whether they are all right.

According to the Kids Clubs Network (formerly the National Out of School Alliance), the national organisation of out-of-school care schemes, approximately one in five children is left home alone during school holidays and one in six goes to an empty home after school. In 1990, the UK only had facilities for 14,000 school-age children. Nearly 2.5 million school children under 16 have working mothers but, even if you believe that the older of these can manage on their own, the current level of available care is a scandal. The Kids Clubs Network has launched an Out of School Childcare Service to help employers introduce care schemes into their companies and has published a booklet entitled *Childcare for School-Age Children: An Employers' and Employees' Guide* (call 0171 247 3009 for further information).

The government is also involved in the expansion of such provision. From 1993–96 the Conservative government spent £45 million to help establish more than 50,000 after-school and holiday childcare places. The initiative was launched to allow mothers of school-age children to take up employment or training opportunities. The Labour government is expanding this scheme by concentrating on after-school homework clubs.

Unfair Childcare Tax Law

The government is using tax law to create incentives for employers to provide childcare for their workers – employers can now obtain tax relief from providing or subsidising childcare. Employees, depending on how much they earn and what kind of childcare their employers provide, may not have to pay taxes on help provided. But the tax law is not even-handed and does benefit the employer more than the employee. Under the general business tax rules employers can claim tax relief if they:

- give an employee a cash allowance for childcare
- pay the fees of a nursery, childminder or nanny
- pay for the running costs of a nursery they provide for employees' children.

But as an employee you have to pay tax on employer financial help with childcare unless

- your child has a place in a nursery provided by your employer (i.e., your employer provides the premises or is wholly or partly responsible for financing and managing the nursery), *or*
- your employer directly engages and pays the fees of a nursery, childminder or nanny *and* you earn less than £8500 a year.

That means that most families still have to pay tax on employer help with childcare. Efforts are being made to get this law changed so that all forms of employer subsidies of childcare are not taxable.

So far, there is only a minor subsidy to help parents on low incomes to pay for childcare costs if the lone parent or family is eligible for Family Credit, Housing Benefit, Council Tax Benefit or Disability Working Allowance. This subsidy is called the 'Childcare Disregard' and allows a percentage of childcare costs to be deducted from earnings when the government is determining whether the family is eligible for such benefits. This subsidy is not useful for most of you however – it only affects about 150,000 women.

MAKING SLOW PROGRESS

As you can see, despite a few initiatives by the national government, most progress is being made through individual initiatives by workers, local authorities and some forward-looking firms. A useful resource to find out about these initiatives is Scarlett McGuire's book *Best Companies for Women* (McGuire 1992), which lists the best 50 companies in the UK as employers for women, using criteria based on the companies' policies on recruitment, training, promotion, equal opportunities, childcare, career breaks, job share, part-time working, and term-time working.

Recent British governments have not taken any real leadership in improving the workplace for women. The then Prime Minister John Major's support of Opportunity 2000 (the business community's effort to improve women's chances in the work world) was, unfortunately, more cosmetic than real. It is the European Union that has acted as a catalyst for change.

Compared with most Western European countries, British policy regarding pregnancy and maternity benefits is far behind. Compared with the United States, however, the only Western industrialised country with no national maternity benefits policy (aside from twelve weeks of unpaid leave), Britain is way ahead.

The Pregnancy Friendly Workplace

While the need for maternity benefits and childcare facilities is acute, the task of achieving a workplace free of reproductive hazards remains the most difficult. There is some legal protection under the 1974 Health and Safety Act and the Management of Health and Safety at Work Regulations 1992, as well as the regulations about specific workplace hazards which implement the European Union Work Related Directives, but the question is how well these laws actually work. Later chapters will discuss these initiatives and the appendices at the end of the book provide more detailed information.

Despite existing legislation, however, women workers who consider becoming pregnant or who are already pregnant continue to face scientific uncertainty about reproductive risks, myths about their capabilities and place in the workforce and strong economic pressures against cleanup costs. The following chapters will help you obtain a more pregnancy friendly workplace.

CHAPTER 1 APPENDIX 1: WOMEN'S LABOUR STATISTICS

Women and men are generally employed in different industries. Women continue to be significantly underrepresented in the primary sector (agriculture, energy and water) in most manufacturing, in transport, in communications and particularly in the construction industry. They remain over-represented in three major occupational groups – clerical/secretarial, personal and protective services, and sales. Fifty-two per cent of all working women are employed in these areas against only 18 per cent of employed men.

Women's participation in the labour force is highest in the middle age groups 35–44 and 45–59. More than half of the women who work part time are over 40 years of age. Part-time jobs in the UK almost doubled from 15 per cent in 1971 to 28 per cent in 1994, the biggest increase being in the public and retail sectors. Not all women wanted full-time jobs; many needed time for their family responsibilities. Yet despite this, during the decade 1986–96, the number of employed women with children under five years of age increased from 40 per cent to 54 per cent. As would be expected, women with higher qualifications were far more likely to be employed. Among women with children under five years of age, 74 per cent of highly qualified mothers were employed compared with 31 per cent of those least qualified.

Where Women Work, 1996
(Percentage female in each category)
75 per cent clerical and secretarial
68 per cent public administration, education and health
53 per cent distributors, hotels and restaurants
46 per cent banking, finance and insurance
32 per cent managers and administrators
27 per cent manufacturing
26 per cent agricultural and fishing
22 per cent transportation and communication
19 per cent energy and water
10 per cent construction

Women's wages are less likely to exceed the minimum rates than men's earnings and, according to the spring 1996 labour force survey, women earned on average 80 per cent of men's hourly wages. Women often get paid less than men for doing virtually the same job; a frequent occurrence even though it is illegal. Companies achieve this by assigning men and women different job titles to disguise what is essentially identical work. A woman, for example, may be classified as a secretary and a man as an administrator.

Not only are women often paid less for doing the same work, but stereotyped 'women's jobs' – such as retail sales, health care and teaching – bring lower wages than jobs filled mainly by men. To correct this pay scale bias, a 1984 amendment to the Equal Pay Act mandated that work which has been rated as equivalent in terms of effort, skill and decision-making should receive the same pay regardless of the sex of the employee. Despite this attempt, women's wages still lag behind.

Part of the reason lies in the characteristics of the labour market. The core sector of the labour market consists of stable, well-paying jobs. The large secondary sector of the labour market is characterised by low wages, poor working conditions and job instability. Some well-educated women have been able to enter the core sector and are making inroads in the traditional high-paying male jobs such as chartered accountant and corporate executive. But women are over-represented in the second category, jobs which tend to be in small firms, filled by women and part-timers with high labour turnover – features which make it difficult for trade unions to organise.

In addition, people in part-time jobs are likely to receive less health and safety protection. According to the Workplace Industrial Relations Survey, in 1992, only 24 per cent of non-unionised employees had health and safety representation, down from 41 per cent in 1984. In 1996, 82 per cent of part-time workers were women.

(The information in this appendix is taken from the *Employment Gazette*, April 1993 and *Labour Market Trends*, March 1997.)

2

What Can Go Wrong: Reproductive Hazards for Man, Woman and Child

LEARNING ABOUT OUR REPRODUCTIVE SELVES

Before we can improve our workplaces, we need to know a little about the biological basis of reproduction and the points of vulnerability from occupational hazards. Reproduction is a complicated business and the course of procreation is not smooth. Many fascinating pieces of the puzzle have been discovered, but still more remain to be unearthed. We particularly want to know what goes wrong and why.

About 10–20 per cent of couples in their childbearing years are infertile. In about half the cases, the male is totally or partially responsible for the difficulties in conceiving. About two-thirds of infertile men suffer from some type of poor sperm function, of which doctors do not know the cause. Even when conception does occur, only about one-quarter to one-third of fertilised human eggs are likely to result in the birth of a baby. Many pregnancies end in miscarriages. About three-quarters of early miscarriages show chromosomal or other abnormalities. In addition some babies are born with congenital malformations, others show developmental lags that appear within the first year, and still others suffer from additional physical and mental disabilities appearing later in childhood or adolescence.

Environmental influence is seen to be of major importance, and society is now putting pressure on pregnant women to give up certain lifestyles that were never questioned in earlier times. Women are told not to drink alcoholic beverages, not to smoke, to limit coffee, not to take medication unless absolutely necessary, to avoid the recreational use of drugs and to be aware of workplace hazards.

Scientists still know woefully little about the reproductive effects of environmental and occupational exposures and until recently researchers barely looked for damage to the male from these types of hazards. As a result of this bias, women were, and still are, assigned the major responsibility for infertility and for any harm to their pregnancies or unborn child. This is a heavy and unwarranted burden to carry. Men, meanwhile, carry the burden of being harmed while remaining blissfully unaware of that fact.

Women, too, are relatively ignorant about male reproductive functions. We are taught from childhood that reproduction is primarily a woman's concern and accept this interpretation as a scientific fact. Even when evidence is to the contrary, women shoulder the reproductive blame at some emotional level. Interviews with infertile couples have indicated that the wife frequently feels like a failure even when the biological problem lies with her husband.

Biological interpretations, however, have differed during various historical periods. They have not only been based on advances made in science but on the views held by society at certain times about the roles of men and women. In the 1600s, for example, the male was thought to play the primary role in reproduction. As strange as it may seem today, one popular hypothesis was that each male sperm contained a complete human being. Textbook drawings showed a little person standing inside the sperm cell. The woman was merely the incubator and was seen as playing only a minor part in determining the characteristics of her child. Now, almost 400 years later, the tables are turned.

Some Known and Suspected Causes of Infertility in Men and Women

- Environmental pollutants
- Workplace toxins
- Infectious diseases
- Genetic defects
- Side-effects of medication
- Inadequate nutrition

- Increased age
- Drugs
- Stress

WHAT CAN GO WRONG

Once we learn how complex the process of human reproduction really is, beginning with egg and sperm cell formation and culminating in the birth of a child, we may simply be amazed that everything turns out right the majority of the time rather than taking the time to wonder what goes wrong in the remaining cases. There are many possibilities. For instance, if the hormone balance is upset in women, ovulation may not take place. Disruption of the hormonal system in men may interfere with sperm production. Direct damage to the testes and sperm-producing cells can result in insufficient or abnormal sperm. Because male and female reproductive organs develop from the same embryonic tissue and by similar developmental pathways, there may be little difference in their sensitivity to mutagens. Exposure to toxic material can reduce sexual desire and damage reproductive functions in both males and females. Such exposure can change genetic material in either the egg or sperm, resulting in a miscarriage or a live born child with birth defects. Changes in genetic material (called mutations) are passed on through generations.

Another area needing further research is how the body repairs genetically damaged cells from which eggs and sperm mature (precursor cells). Sometimes the body cannot repair these cells correctly. This error in the repair mechanism can cause mutations. Little is known about this process, but so far there is no reason to assume that male sperm precursor cells are more capable of repairing genetic damage than female egg precursor cells.

So, what are some of the essential differences in the way eggs and sperm are produced? Each month an ovary of a healthy woman normally releases one egg. A female baby is born with undeveloped potential eggs already in her ovaries. These are called primordial germ cells and remain dormant in what is called the oocyte stage until puberty. Sometimes, it is only then that toxic harm from the prenatal stage of development becomes known. Women never generate any more oocytes and therefore a substance toxic to oocytes damages a finite supply.

Table 2.1 A comparison of female egg cells and male sperm cells

Males	*Females*
1. New sperm cells are generated continuously from the age of puberty.	1. Eggs are produced from precursor cells that are formed in the embryo.
2. Cells in testes from which sperm cells mature are present from birth and can be cumulatively exposed to mutagens.	2. Eggs mature cyclically and are released by the ovaries approximately every 28 days.
3. Sperm mature cyclically. They start developing at close intervals and run concurrently, taking 70–80 days to develop. Rapidly increasing cells are generally more sensitive to genetic damage than cells that grow slower.	3. Egg precursor cells are embedded in layers of connective tissue in the ovaries within the body.

While the female has a finite amount of potential eggs, they are better protected than the sperm from toxic harm as they are embedded within the ovaries. The male continues to regenerate spermatogonia (the male counterpart of the oocyte) after puberty. Sperm cells take 70–80 days to mature and are continuously produced from these spermatogonia. As this process takes place near the surface of the body, the developing sperm are more exposed to toxic injury than the unreleased eggs. Unless an exposure is continuous or there is chromosomal or hormonal damage affecting the growth of normal sperm, a toxic dose over a brief period of time will only hurt approximately a three-month supply of sperm. Completely fresh cells that were not exposed to the toxic substance are then generated – this is probably nature's way of balancing things out. The woman cannot replace her eggs once they have been damaged, but they are fairly well protected. The man can replenish his sperm supply, but it is more vulnerable to short-term toxic exposure.

Even if the egg and sperm remain healthy, there is still a long road ahead to conception, let alone the birth of a child. Only a few sperm ever reach the egg. For example, assume that the number of sperm reaching the ovum (egg) is probably less than 1 per 100,000 ejaculated by the male. This means that somewhere between 10–100 sperm actually come close to the egg. We do not know what, if anything, is special about these particular sperm. Are they just a random sample of all the sperm ejaculated at one time or are they a special group of sperm cells selected by the female's body on the basis of as yet unknown criteria?

The effect of exposure to toxic substances on the fertilised egg's ability to implant itself in the womb is also unclear. Immediately after conception, any interference with the cell division of the fertilised egg or its implantation in the uterine wall causes the embryo (as the unborn child is called during the first few weeks after conception) to die. The ensuing miscarriage is so early that women only know that their menstrual period is a few days late or heavier than usual. If the embryo survives the early days it has passed its first hurdle, but there are more to come.

The embryo, and later the foetus, is usually more sensitive to toxic substances than the mother. This can lead to problems as most substances present in the mother's blood stream whether they entered through her digestive system, were inhaled or absorbed through the skin, pass through the placenta to her unborn child. The first three months of pregnancy are an extremely sensitive period as during this time the organs are formed and developed, and major deformities in the unborn baby's heart, brain, limbs or other organs can occur. Substances causing these types of defects are called teratogens. They can cause devastating foetal damage during a specific pregnancy but are neither transmitted to future generations nor likely to occur in another pregnancy unless the mother is again exposed to the hazardous substance.

During the second and third trimesters the foetus develops and grows. Childhood developmental problems can originate during this period. Anything that causes the baby to be born prematurely or have a low birth weight, such as stress, toxic exposures or poor nutrition, puts it at more risk for illness and death.

Checking Reproductive Health

Female Health

1. Personal, medical and family history
2. Physical examination
3. Functioning of the ovaries
4. Secondary sex characteristics
5. Receptivity to sperm of cervical mucous
6. Monthly thickening of the uterine wall (endometrial cells)
7. Condition of fallopian tubes and uterus

Male Health

1. Personal, medical and family history
2. Physical examination
3. Semen quality
   a) appearance of semen
   b) pH (degree of acidity) of semen
   c) amount of semen ejaculated
   d) number of sperm per millilitre of semen
   e) sperm movement patterns
   f) ratio of live/dead sperm in semen
   g) sperm shape and size
   h) ability to penetrate cervical mucous
   i) sperm–egg interaction

Source: Adapted from material in US Congress, Office of Technology Assessment (1985) *Reproductive Health Hazards in the Workplace*, US Government Printing Office, Washington, December, Chapter 5.

Some cancer-causing agents can cross the placenta, thus causing direct reproductive damage to the foetus. While the placenta acts as a partial barrier, it by no means filters out all potential toxins. Because the lag between exposure to a carcinogen and the appearance of cancer can range from 5–40 years, many children who were exposed as foetuses may develop cancer many years later and they and their parents never realise the connection.

So, reproductive problems occur at every step of the process and, as a result, toxic substances can cause reproductive harm at all stages – prior to conception, during pregnancy and (as will be discussed at the end of this chapter) after birth. Toxins can affect menstruation, ovulation, sperm production, sperm quality as well as cause miscarriages, birth defects and cancer.

Your Body Changes in Pregnancy

Understanding the biological changes in your body during pregnancy and learning how your workplace environment interacts with these body changes will enable you to be healthier and more comfortable at work. For example, if you suffer from neck, back, and eye strain from sitting in front of a video display unit (VDU) all day, you may lose your appetite and not eat enough to provide sufficient nourishment for yourself. But, if you are pregnant, you may also not eat enough to provide adequate nourishment for your growing foetus. Most important, you need to improve the fit between your workplace and your pregnancy.

Most pregnant women are aware of the obvious anatomical, physiological and psychological changes during pregnancy. In early pregnancy, fatigue and nausea often affect comfort and efficiency. In late pregnancy, a protruding abdomen, increased weight and retention of fluid can make walking or changing positions awkward and take some of the pleasure out of any job. Rapid mood changes and exaggerated emotional reactions may also occur. We seldom know, however, about more subtle changes, or even the reasons for the more obvious changes. We frequently realise that poor workplace conditions aggravate discomfort but not how proper facilities can alleviate it.

- Major physiological changes occur in the circulatory, respiratory, muscle–skeletal and endocrine systems. The volume of blood increases 30–40 per cent. This increase is largely composed of plasma and red blood cells. Higher heart rates are normal during pregnancy and they rise more than usual when participating in strenuous activity. You may be particularly prone to this increase if you were anaemic, inactive, or very overweight prior to pregnancy. Furthermore, blood sugar levels are naturally lower during pregnancy and pregnant women tend to burn carbohydrates faster. This combination creates the risk of an abnormally low blood sugar level during periods of heavy activity.
- During pregnancy, you may hyperventilate. This means that you continually move more air through your pulmonary system to extract a given amount of oxygen than when you are not pregnant. This hyperventilation is due to the elevation of the diaphragm by the uterus and by a 20–30 per cent increase in your consumption of oxygen over the non-pregnancy level.
- During pregnancy you have to adjust to a different level of hormones. The hormones produced by the pregnancy cause ligaments, tendons and other connective tissues to soften, making you more prone to injury if you carry out tasks that excessively stress or stretch your joints. The expanding uterus and loosened

joints put increased stress on your back during pregnancy and it remains sensitive to injury for several months after delivery.

- In the later months of pregnancy you may fall more easily; the body's centre of gravity shifts and throws you slightly off balance. At this time, moreover, the enlarging uterus may press on several organs. It can press on the sciatic nerve, making your legs buckle, or it can press on the large vein carrying blood from the lower parts of our bodies, causing your legs to swell. Constipation, haemorrhoids and varicose veins can cause discomfort, and the baby's head deep within the pelvis may press on the base of the bladder causing frequent urination. You can suffer from fatigue, laboured or difficult breathing, and insomnia resulting from the combination of the effects of weight gain, increased respiratory requirements and the discomfort of manoeuvring a more bulky and awkward body.

Modifying Your Workplace to Accommodate These Changes

Once you are aware of these body changes, you need to know how the workplace can be modified to ensure you greater comfort. A comfortable workplace depends primarily on good design of work space and equipment and a reduction in physical and social stress. Modifications that improve conditions for a pregnant worker also make the workplace healthier and more comfortable for all workers of both sexes and lead to greater efficiency and a lower absenteeism rate. These points are worth emphasising to co-workers and employers so they do not believe that pregnant workers are receiving preferential treatment at their expense. Another way of adapting body changes to work demands is by trying to arrange an alternative work schedule. Flexible arrangements can help you to adapt to physiological changes in your body while maximising your efficiency on the job.

If morning sickness is a problem for you, then a schedule starting later in the day can alleviate the tension between your work and your pregnancy. If fatigue is a problem, then a reduced work-week may be an answer. If combining household chores and errands proves difficult, a nine-hour, four-day working week may be your salvation. Your stress might be reduced if you could concentrate on your paid work during these four days, you would have one weekday to take care of household tasks, shopping and health care needs and more quality time to spend with your family and friends over the weekend. You could receive all these stress-reducing benefits without loss of income or fringe benefits, and without putting your job in jeopardy by excessive absenteeism or putting your pregnancy in jeopardy by working a schedule that stretches your physiological and psychological capabilities to their limits. These possibilities were discussed in more detail in Chapter 1.

BRINGING THE MALE BACK IN

Prospective parents can find many books and articles about women and what can go wrong with pregnancy, but few have been written about men and what can go wrong with pregnancy. Not only has the 'little person standing inside the sperm cell' disappeared from these books, but so, too often, has the big person who is generating the sperm. Because the emphasis in reproductive research has focused on the woman's role, the potential for birth defects arising from the result of the father's exposure to chemical substances is largely unknown. On the basis of current understanding of human reproduction, however, such birth defects could occur in three ways:

1. by gene mutation
2. by chromosomal mutation
3. by acting as a teratogen (interfering with normal embryonic or foetal development) carried in the semen during intercourse.

Some scientists think that toxic substances absorbed by the exposed male may contaminate the seminal fluid and cross the placental barrier through intercourse. They then can be absorbed by the fertilised egg causing miscarriages or birth defects. A few studies and case reports suggest that this may indeed happen. One report describes a woman who bore three children with severe malformations over a period of years when her husband was suffering from lead poisoning. Subsequently all three children died. Pregnancies before her husband had developed lead poisoning and after he had recovered led to the birth of normal children.

As far back as 1934, Dr Alice Hamilton believed that lead could affect sperm cells. She based this opinion on a Japanese study reporting that where the husband had been exposed to lead, twice as many couples were childless as would be expected. A 1985 review of scientific papers analysing the effects of lead on human health also revealed an unusually high number of abnormal pregnancies among wives of lead workers. Although there is increased evidence that these abnormalities are probably due to the male's exposure to lead, there have so far been no conclusive studies proving or disproving this.

Male employees exposed to anaesthetic gases have also been studied. In the late 1970s, a committee of the American Society of Anesthesiologists (ASA) sponsored by the National Institute of Occupational Safety and Health (NIOSH) conducted a nationwide study of operating room personnel. The inquiry found that the risk of exposed male physician anaesthetists to having offspring with birth defects was 1¼ times that of unexposed medical personnel. Another ASA study of dental surgeons who were exposed to anaesthetic gases for periods exceeding three hours per week found an increase of miscarriages among their wives. An English study of operating room personnel in the 1970s, however, did not find an increase in the miscarriage rate among wives of exposed males but did find a higher rate for certain birth defects in children of exposed fathers.

Research findings have differed. One of the problems posed by investigations based on workers' memories of past events (retrospective studies) is that there may be biased recall of both the extent of exposure and exact reproductive injury.

The list of agents suspected of causing reproductive damage through male exposure keeps getting longer. Men who work in factories which produce plastic products and fire fighters who come into contact with burning plastic material often inhale vinyl chloride (VC). Vinyl chloride can affect genetic material in sperm. Studies in four countries have found that workers exposed to polyvinyl chloride (PVC), one of the most commonly used plastics, have an excess of chromosomal changes. Wives of workers exposed to vinyl chloride have been found to have higher than expected miscarriage and stillbirth rates.

Ionising radiation, carbon disulphide, oestrogen, DBCB and extreme heat are other villains. Ionising radiation can kill sperm-producing cells of the male testes. A study of Japanese radiological technicians showed that both males and females had a higher sterility rate than the general population. Research on male workers exposed to carbon disulphide for three years (used as an insecticide, solvent and in the manufacture of viscose rayon) found that the exposed men had decreased sex drive and trouble in having an erection. They also had five times greater rates of sperm abnormalities than the control group not exposed to the chemical.

Several pesticides and herbicides have been linked to male reproductive organ dysfunction and to birth defects in the offspring of exposed male workers. In one report four out of five members of a farm working crew exposed periodically to a wide variety of pesticides complained of impotence. Accumulating data revealing that Dinoseb, which had been sprayed on crops in the United States for more than 40 years, could cause birth defects and sterility as well as cancer, cataracts and immune system damage finally became persuasive enough for the United States Environmental Protection Agency (EPA) to use its emergency powers to order an immediate ban in October 1986. The only two other times that the EPA used its emergency powers was to ban the agricultural chemicals 2,4,5-T, an ingredient of Agent Orange and ethylene dibromide (EDB), a strong human carcinogen which was suspected of damaging both the quality and quantity of exposed men's sperm.

To add further to the complexity, some substances can have several effects. For example, the pesticide DBCP is a mutagen which can cause direct damage to the chromosomes. It can also destroy the cells in the testes that are responsible for producing sperm (the male germinal epithelium) and thus cause male sterility. In 1977 semen samples collected from 308 American workers exposed to DBCP in their jobs found that 50 per cent of these men had either no sperm or a very low sperm count in their semen, nearly twice the rate expected for the general population. Impotence, as well as sensitive and enlarged breasts, are symptoms of men who produce oestrogens, oestrogen-based drugs, DES pellets and pastes and administer DES to animals.

Paternal exposure to electromagnetic energy is also suspected of being a contributing factor to their offspring being born with either a club foot or Down's syndrome. In addition, mutagenic and carcinogenic effects have been reported. This is quite worrying, as advances in magnetic and superconductor technology means that even more workers will be exposed to weak electromagnetic fields. Researchers in Europe and the United States are currently investigating the possible links between electromagnetic fields and harm to human reproduction.

Because so little emphasis has been placed on male reproductive hazards for so long, both the development of research studies in this area and the distribution of new research findings seem very slow. Even when an agent is suspected of being a reproductive hazard the information is not always forwarded to the employer, and even when the employer knows, he or she may deliberately neglect to pass this information on to employees on the grounds that it has not yet been proved.

Table 2.2 Work exposures associated with abnormal human male function

Confirmed Effect	*Inconclusive Effect*	*No Observed Effect*
carbon disulphide	anaesthetics	epichlorohydrin
DBCP	arsenic	glycerine
lead	benzene	P-TBBA
oral contraceptives	boron	PBB
	cadmium	PCP
	carbaryl chlordecone (kepone)	
	DNT and TDA	
	ethylene dibromide	
	manganese	
	mercury	
	pesticides	
	PCP	
	radiation, ionising	
	radiation, non-ionising	
	solvents	
	TCDD (dioxin)	
	vinyl chloride	

Source: Adapted from Susan D. Schrag and Robert Dixon (1986) *WOHRC News*, vol. 8, no. 2.1, Women's Occupational Health Resource Centre, School of Public Health, Columbia University, New York.

Take the case of Ken, a forker in a heating and plumbing manufacturing firm. It was his job to fire pieces in the furnace to prepare them for enamel finishing. His employer never informed the men working in the furnace that the intense heat could damage their sperm count. It was only when it was too late that Ken learned the truth.

> I didn't realise there was a problem until the doctor told me my sperm count was too low to impregnate my wife. He told me that he was 90 per cent sure my sperm count was irreversibly low because of the intense heat at my job. There is no medical treatment available to correct my problem.
>
> I was angry enough to take the company to court, but I didn't because my doctor told me it would be too hard to prove my low sperm count was caused by the heat work. He said they could say that taking hot showers also caused low sperm counts.
>
> Ken, US furnace forker

Table 2.3 Some substances known or suspected of harming male reproductive health or the health of their offspring

Substance	*Effect*
Lead used in making storage batteries and paints	Fewer sperm, sperm that move more slowly than normal (decreased sperm motility), and more oddly-shaped sperm (increased malformation of sperm)
DBCP (dibromochloropropane) a soil fumigant now banned	Mutagen, lowered sperm count, testicular dysfunction
Ionising and non-ionising radiation, the former found in nuclear plants and medical facilities, the latter in high voltage switchyards and in communications facilities	Possible damage to germ cells and lowered fertility
Anaesthetic gases	Unexposed female partners are thought to have higher than normal number of miscarriages
Vinyl chloride used in plastic manufacturing	Unexposed female partners are thought to have more miscarriages and stillbirths
Kepone used as a pesticide	Possible loss of sex drive, lowered sperm count and slower movement of sperm
Heat stress occurring in foundries, smelters, bakeries and farm work	Lower sperm counts and sterility
Carbon disulphide used in the manufacture of viscose rayon and as a fumigant	Possible loss of sex drive, impotence and abnormal sperm
Oestrogen used in the manufacturing of oral contraceptives	Possible loss of sex drive and enlarged and sore breasts
Methylene chloride used as a solvent in paint strippers	Possible very low sperm counts and shrunken testicles
EDB (ethylene bromide) used as an ingredient in leaded gasoline and as a fumigant on tropical fruit for export	Possible lower sperm count and decreased fertility in wives of workers

Two other forkers that Ken knew about also had fertility problems. One was permanently sterile and the other, who had only been doing the job for a year, eventually had children after transferring to another department. Despite these known cases, the company has never changed working conditions to reduce the potential risk of infertility nor informed workers of this possibility.

In Ken's case the probable cause of his infertility was easy to determine, but long time lags between exposure and reproductive damage sometimes make it difficult to establish cause and effect. This is why the possibility of a link between paternal workplace exposure and community exposure to low levels of radiation from the Sellafield nuclear reprocessing plant has caused such a furore. Ever since the discovery in the early 1980s of more than the expected number of children with leukaemia in this region, the connection between radiation and childhood cancer has been raised. In the United States, the March of Dimes (an organisation concerned with the eradication of birth defects) is also honing in on the father's role in reproductive damage. They are looking into the possible effects of workplace and environmental pollution. When scientists were focusing only on the maternal/embryo effects, they were ignoring the fact that the male reproductive tract contains very metabolically active cells vulnerable to poisons and mutagens.

When clusters of childhood cancers, miscarriages or birth defects occur, they always should be seriously investigated. But occasionally fate plays its hand and an extraordinary event can affect a very safe job.

> My job was to administer and support English language programs overseas. It is mainly an office job combined with visits to the overseas classes in the program. Several years ago, I was on an on-site visit near the Polish border with Russia when the Chernobyl nuclear accident occurred. I was tested for radioactivity and the soles of my feet were contaminated. This lasted for a week. I never had any follow up, but so far I don't seem to have any side effects.
>
> William, supervisor of overseas English language programmes

MEN AND THEIR PERCEIVED REPRODUCTIVE VULNERABILITY

While men are very aware of the risk of developing cancer from occupational toxins, they focus less on reproductive harm despite evidence linking their fertility and their partners' miscarriages to their own workplace exposure. Part of the denial of a male role is due to society's narrow interpretation of reproductive harm – one limited to the effects on the pregnant woman and unborn child. As was apparent in a few of the interviews, machismo also contributes; some men do not like to talk about any difficulties remotely concerned with sexual adequacy and prowess. When asked about whether he ever lost his sex drive, became impotent, or had a low sperm count, one of the men responded:

> Now look, between my wife's headaches and backaches and having to get up early and things like that you can't afford to lose no sex drive. You better get it while it's hot.
>
> Robert, US plastics fabricator

Even when men are aware of the hazards, they do not seem to vocalise their concerns as much as women. A few interviews showed that men are more likely than women to accept the reassurances of their companies and unions. Take the case of Alan, an American firefighter. Firefighting is one of the most hazardous jobs in terms of exposure to all sorts of highly dangerous toxins. Despite the progress firefighters' unions have made in providing information, firefighters still do not know what hazardous substances they are facing in most fires. Yet Alan, outwardly at least, remains reassured.

> In any fire there will be gases from the smoke. When different plastics burn they give off poisonous smoke. It all depends on who has the fire. The closest I came in contact with poisonous gases was in one house fire where there was a refrigerator smouldering. I guess it was insulation on the inside so we pushed that outside the door without our masks on and all of us were sick for a week.
>
> As long as we're aware of what is on the premises, we can do everything possible to prevent any harm to our offspring from our exposure. But as long as we're not sure what different chemicals are in the building, there is no way for us to be totally protected. The firemen's national organisation is always doing something. There is always some one in the national, state or county federations that is working on the problem and trying to get everything that we need to make our job much safer. Unlike a lot of jobs, information doesn't filter down as slowly with us because we're a national organisation

> and the information is usually distributed rather quickly. How fast somebody might act on it from city to city may vary and getting the right equipment takes more time and money.
>
> I am under a lot of stress in my job and at times my sex drive is increased. I think it is primarily to relieve tension, because afterward you feel rather relaxed. You know how it is, don't you?
>
> Alan, US firefighter

Male workers operating in other types of toxic environments also voice concerns which are combined with either a general satisfaction with safety precautions or a refusal to think about the implications of the hazards.

> We come in contact with a lot of chemicals such as ink solvents and resins. Resins are a powdery substance that would probably have a danger of going into your lungs. We also use some acids and a couple of dangerous gases, hydrogen, nitrogen and things like that. But we use a lot of precautions.
>
> I had an opportunity to work with machinery that has radiation, but I was not exposed to it. There was a chance of inhaling the gases. I never inhaled it. We work for a union company and everyone is concerned with strict safety. We work with a mask, proper equipment, goggles and gloves. You work with these substances but don't come in contact with them. They have safety inspectors come in periodically to see if everybody is working safely. It is very dangerous, a lot of bad things could happen.
>
> I can see something happening to myself or my children if sometimes I got careless or I got into a situation where I was exposed to the chemicals. I am very conscious of that so I take precautions. I don't see it really happening.
>
> Larry, US lab technician

> We smooth out plastic parts to be put in the inside of aeroplanes. Sometimes we paint them. I come in contact with different paints and there is dust all over from sanding the plastic. Sometimes I'm pulled out of where I'm working and have to go downstairs where they are doing all the spray painting. Even to go to the bathroom, you have to go through the painting section. Downstairs the air was better because they had the doors open. Upstairs there were only a few windows.

> Over the last year or so I've become more conscious of the dust in the air. I figure it's uncomfortable thinking about it so I turn off about it. But the safety factor keeps sticking in my head. It's hard to say in a factory like this what the other guys and their families think. Guys take off sick and they may go fishing or something like that. We're a close group of guys, but we don't really know anything about each other.
>
> My company is a good company. I'm not trying to bad mouth the company. But I figure the dust is just something that nobody paid too much attention to.
>
> Robert, US plastics fabricator

The male workers express their concern, try to be extra careful and tend to feel successful in their efforts, whereas the women remain more uncertain.

TOXIC EXPOSURE AFTER BIRTH

In addition to increased awareness about male reproductive workplace hazards, recognition of workplace hazards affecting infants and children is also growing. Just when you can breathe a sigh of relief that you had a healthy pregnancy and that your baby has been born without any birth defects, new worries emerge. The two most common avenues of toxic exposure to a child after it is born are from dust brought home on the parents' work clothes and through breastmilk. For example, because oestrogen can be absorbed through the skin or inhaled, young sons and daughters exposed to oestrogen dust brought home on their mothers' or fathers' clothing develop sore and enlarged breasts which disappear when exposure to oestrogen ceases.

Men working with asbestos bring home asbestos dust on their clothing, affecting their wives' and children's health. In Manville, New Jersey, USA (home of the Johns Manville company which claimed bankruptcy rather than be liable for millions of dollars of health damage claims against the company), research physicians are studying the possible health effects on the third generation – the grandchildren of the exposed employees.

Breastmilk pollution is a newer problem. Recently, the World Health Organisation presented worrying evidence that toxins, entering mainly through the route of the food chain, are now found in women's breastmilk in most areas of the world. How serious is the problem? Does the infant suffer any ill effects?

Breastmilk Pollution

A nursing mother typically produces a litre of milk per day providing her baby with nature's best nourishment. Unfortunately in today's contaminated atmosphere, breastmilk can also pass on a wide variety of substances present in the mother's tissues or blood. Chemicals or drugs can bind to milk protein or to the surface of milk fat globules and are included in the breastmilk along with protein, fat, carbohydrates, minerals, vitamins, hormones, and antibodies.

Scientists studying the breastmilk path of possible health effects on the nursing infant assume the existence of a dose response. That means that a high concentration of a toxic chemical found in breastmilk might harm the baby but a small amount of the same toxic chemical probably would not. They are looking into acute short-term effects, long-term chronic effects due to the build up of small amounts of toxic agents over time and the interactive effect between chemicals. Sometimes a chemical is not harmful by itself but becomes harmful if another chemical with which it interacts is also present.

Accidental chemical spills in several countries, rather than much lower dose occupational exposures, sparked concern about breastmilk chemicals. (Discovering the AIDS virus in breastmilk was a further disturbing finding.) In Japan, a faulty production process resulted in PCBs (polychlorinated biphenyls) getting into rice cooking oil. Some unborn babies were exposed while in the uterus and this was further complicated by PCBs reaching the baby through their mother's milk. Similar incidents occurred involving acute methyl mercury poisoning. People ate contaminated fish in Japan and contaminated grain in Iraq, with one of the consequences being that pregnant women and nursing mothers developed high mercury blood and milk concentrates which affected foetuses and nursing infants leaving them with damaged central nervous systems and suffering from a disease resembling cerebral palsy.

In Michigan, PCB-contaminated fish and cows fed with PBB-contaminated grain resulted in high levels of these chemicals being present in the food chain. Even though high amounts of PCBs and PBBs (polybrominated biphenyls) were found in the milk of nursing mothers, their babies did not seem to suffer any ill effects. In Hawaii, the pesticide heptachlor sprayed on pineapple tops (which had been taken off the market for all other purposes because it was considered to be too toxic to humans) also found its way into the food chain and eventually into breastmilk and the nursing infants. This raised quite an uproar, and heptachlor was finally banned from use on the Hawaiian pineapple crop. In 1989, government scientists investigating chemical contamination of food and the environment in Britain discovered traces of dioxin in women's breastmilk. This is alarming enough, but more upsetting was that some samples were well over the recommended safe limit.

Ideally a mother planning to nurse her baby should not be exposed to mercury, lead, solvents, many pesticides, and PCBs and PBBs in the workplace either before she attempts to conceive, during pregnancy, or during lactation. She should also try to avoid exposure to anaesthetic agents, cadmium and pharmaceutical agents, particularly oestrogens and chemicals involved in the production of viscose rayon and synthetic rubber. Women who are self-employed, work as household helps, use chemicals in hobbies such as arts and crafts or engage in home improvements should also be alert to possible contaminants of their breastmilk.

Past maternal exposure produces a body burden which is not readily excreted in faeces or urine. Instead, chemicals are retained in the mother's tissues and bones and are released when she begins to nurse. Chemicals that occur at higher concentrations in milk than in the mother's blood plasma are the most worrying; even if the transfer of the chemicals from the mother's blood to her milk is extremely low, a high concentration can accrue in the milk because the blood flow to the mammary tissue exceeds milk production by 400–500 times.

During lactation, transfer to milk occurs readily for fat soluble substances due to the high fat content of breastmilk. It is likely that fat-soluble chemicals such as DDT, PCB and many other pesticides may be trapped entirely within the milk-fat globule – not a pleasant thought. The amount of PCBs, heptachlor and other similar toxins in the mother's body is determined by her long-term exposure. For example, the pesticide DDT has been banned by the US Environmental Protection Agency for years, yet it is still present in the food chain. There is, therefore, no dietary programme that can be followed during pregnancy and lactation that will eliminate these hazards, though changes may reduce current exposure. Moreover, there are no specific tests to determine the possible effects on the nursing babies. These chemicals are difficult to metabolise or release except in the milk of nursing mothers. PCB levels, however, seem to diminish over the course of breastfeeding and number of pregnancies, so if you have many children and breastfeed for a long period of time, which is not very typical, your younger children may receive less PCBs in their milk.

It is important to remember that so far there have been no documented cases of babies being harmed from being nursed by mothers occupationally exposed to chemical toxins, although blood concentrations of toxic chemicals in occupationally exposed workers typically exceed those in the normal population from 10 to 300 times. There is, however, a recorded incident in the United States involving a mother who regularly nursed her infant while visiting her husband in a dry cleaning establishment. Her baby developed jaundice as her breastmilk was found to contain perchlorethylene.

In addition to the well-founded concern about increases in the total toxic load carried by your body, you should be particularly worried about nursing if you think you are occupationally exposed to high levels of any chemicals that are:

- not water soluble (PBBs, and PCBs)
- not readily metabolised by your body (lead)
- stored in your body and slow to clear from your milk (lead, heptachlor, DDT).

Even if you do not believe that you are exposed to any of the above chemicals, but remain apprehensive about your workplace exposure, find out the chemical names of the substances that you are working with and tell your general practitioner or midwife. Occupational health departments in universities and hospitals may be able to give them the latest research findings in this rapidly developing area of research, and you can then make an informed decision about breastfeeding.

Should I Breastfeed My Baby?

Despite these concerns, the advantages of breastmilk are still seen to be overwhelming because of the protection it provides against infection and the strengthening of the immune system. Doctors recommend that you definitely continue breastfeeding your child except for in cases of extreme contamination.

What You Can Do

The following suggestions may help you to prepare to breastfeed your baby with some peace of mind:

1. Avoid eating fish found in polluted waters. Ask your GP or midwife if they can find out – or advise you where to find out – which fish are considered safe to eat regularly, occasionally or to avoid altogether.
2. Avoid rapid weight loss as this will more rapidly release the toxins into your blood stream and then into your breastmilk. Do not embark on a rigorous diet until after you have stopped nursing.
3. Change your diet. Eat less meat, fish and dairy products. Chlorinated hydrocarbons are more fat than water soluble and therefore tend to remain in the food chain, especially in meat, dairy and fish products. This will reduce your current exposure but will not have an effect on the cumulative toxins already in your system.
4. Request a job change while you are pregnant and breastfeeding if you believe that you are occupationally exposed to a possible reproductive hazard. Electrical, chemical and auto industries and agricultural work are particularly hazardous. You have legal rights set out in the Management of Health and Safety at Work Regulations 1992 (see Appendix C at the back of the book for further information).
5. When you are renting or buying a home, do not choose one that is located near a waste dump or incinerator.

WHERE WE STAND NOW

Preliminary findings indicate that parental (both mother and father) exposure to environmental and occupational toxins is a prime suspect in causing some cases of infertility, miscarriages, reproductive abnormality of both sexes, loss of sex drive, menstrual irregularities, lack of ovulation, male impotence, birth defects, childhood and adult cancers, mental and physical developmental problems and hereditary defects that may only appear in future generations.

But on many issues that are of concern to the pregnant worker, evidence has not yet been found. For example, how much of a hazard, if any, are known toxins in breastmilk to breastfeeding infants? If they are a hazard, which toxins are the most harmful and at what level will harm occur? Will occupational exposure be likely to be severe enough for this dangerous level to be reached? Will general environmental exposure become so high that a small additional exposure in the workplace is enough to tip the scale? Also to date there has been no absolutely conclusive proof, only strong suspicions, that men exposed to occupational reproductive hazards can pass the effects of this exposure directly to their offspring. These suspicions have been bolstered, however, from evidence that the risk for fathering babies with a birth defect caused by a fresh dominant mutation (one that will show an effect even if the child inherits one normal gene) increases with advancing paternal age.

Formerly, it was thought that advancing age only affected egg cells, not sperm cells. It was assumed that if an additional or abnormal chromosome was found it came from the mother. Now studies indicate that around 5 per cent of babies born with Down's syndrome receive the extra chromosome which causes this form of learning disability from their fathers. While this number is small, it is an important finding as the accumulation of data about the male's reproductive risks means that it becomes difficult to place all the blame for birth defects on the woman. In general as men and women grow older they have been exposed to more toxic substances in the workplace, at home and in the community and these exposures may be responsible for at least a small part of the connection found between advanced paternal age and Down's syndrome.

Following the discovery in the early 1980s of clusters of childhood leukaemia near the Sellafield nuclear reprocessing plant in Cumbria, some scientists advocated large-scale research studies aimed at studying whether there is a link between radiation and childhood cancer. In January 1992, the Leukaemia Research Fund and the nuclear and electricity industry funded a five year study of 2000 children up to the age of 15, half of whom were leukaemia sufferers. For the first time, parents of children who have developed cancers will be questioned about their lifestyles, work, illnesses, treatments and the radon levels of their houses. It is known that children of women exposed to X-rays or toxic chemicals such as benzene have a higher than normal risk of developing leukaemia and questions have been raised about the effect of low levels of radiation on sperm. It was hoped that the study will show how childhood leukaemia could be prevented and document the possible health hazards to the population living near the Sellafield plant.

Meanwhile, in 1994 a group of cancer epidemiologists at Oxford University headed by Sir Richard Doll presented evidence that seemed to disprove the theory that fathers' exposure to radiation was linked to childhood leukaemia in their offspring. If this conclusion holds up in additional studies, then more credence will be given to the leukaemia virus theory. As of now, the issue is far from settled.

Many women born in the baby boom of the post-World War II years have already put off having children until they are in their thirties. They cannot postpone their childbearing long enough for definitive evidence to be gathered about male reproductive hazards in the workplace or breastmilk pollution. They have to protect themselves as well as they can based on the available data. Blaming the victim, however, by trying to assign responsibility to the female or male partner, is the wrong way of looking at risks to reproductive health. Expending energy on cleaning up your workplace and environment for both sexes is far more productive. Chapter 3 will go into suspected reproductive hazards of workplaces and the home in greater detail and suggest ways of reducing them. Many women have taken the lead in this area and some of their initiatives are described in Chapter 5.

CHAPTER 2 APPENDIX 1: THE THREE MAIN PRENATAL DIAGNOSTIC TECHNIQUES

1. Ultrasound

Ultrasound analyses embryo/foetal development. It relies on differences in acoustic densities to form images of the unborn child. During the second trimester, it can detect major foetal malformations (such as loss of limbs in the case of the thalidomide babies), multiple pregnancies, and progress of foetal growth. In later stages it can monitor foetal breathing, trunk and limb movement and quantity of amniotic fluid.

There are three levels of ultrasound. Level 1 provides a general picture of the foetus – its overall well-being and gestational age. Level 2 is more sophisticated and looks at specific birth defects such as hydrocephaly (brain filled with fluid), microcephaly (a smaller than normal size brain), neural tube and heart defects. Though birth defects are rare, these are the more common ones and could be related to occupational and environmental exposure. A Level 2 ultrasound should pick up structural defects that might be due to workplace exposure to chemical substances, radiation and infections. Level 3 ultrasound is used for foetal surgery.

2. Amniocentesis

Amniocentesis takes place in the 15th or 16th week of pregnancy and consists of inserting a needle through the mother's abdomen into the amniotic cavity and extracting amniotic fluid. The amniotic sac is the fluid-filled cavity that surrounds the developing foetus. The amniotic fluid contains some live cells shed by the foetus, and both the fluid itself and the cells within it can be analysed to gain important diagnostic information about the foetus.

Amniocentesis is used mainly to diagnose chromosomal abnormalities such as Down's syndrome and other genetic disorders. The amniotic fluid can also be tested for certain enzyme and protein abnormalities; the most common analysis is for alpha fetoprotein (AFP). Abnormally high levels of AFP are associated with neural tube defects such as spina bifida (opening in the spinal column) and anencephaly (lack of most of or the whole brain) and unusually low levels are associated with Down's syndrome.

3. Chorionic Villus Sampling (CVS)

Chorionic villus biopsy is the only method for diagnosing genetic disorders that can be performed in the first trimester of pregnancy. The chorion is the membrane that encases the amniotic sac containing the developing foetus and it is derived from the foetal cells. Therefore, the cells from the chorion are genetically identical to the foetal cells. Analysis of chorionic tissue provides much of the same information as amniotic fluid and cells (except for alpha fetoprotein) but much earlier in the pregnancy.

CVS is a relatively new technique and research has been carried out worldwide determining its effectiveness and safety. Because there is a high incidence of spontaneous abortions during the first three months, it is hard to detect which miscarriages are due to normal early foetal loss and which to the procedure itself. Recent studies, however, find the rate of miscarriage to be comparable with that from amniocentesis if carried out by physicians well trained in the technique. The prospect of a first trimester diagnostic technique is appealing to many women. One of the main drawbacks of amniocentesis is that it can only be done in the second trimester, after the mother feels life and has bonded with her unborn child.

A few other blood tests are available for specific exposures. New blood assays continue to be developed and additional prenatal diagnostic tests will become available. If you are concerned about inherited conditions, ask your midwife to contact a genetic counselling clinic on your behalf.

Currently, prenatal diagnostic techniques cannot detect the effect of most occupational and environmental exposures. Not enough is known about what to look for. For example, most scientists believe that exposure to occupational and environmental toxic substances are more likely to cause miscarriages or developmental problems in the child, rather than birth defects per se. So many factors can cause these occurrences that it is extremely difficult to pin the blame on any one or even a combination of substances. As more data is obtained, this picture will change.

Promising research is being conducted on the harmful interaction of certain genetic characteristics and particular industrial chemicals. Carrier and prenatal diagnostic tests for 'susceptible' workers are likely to follow. These breakthroughs will then likely lead to the identification and counselling of women who are genetically more susceptible to specific industrial chemicals. Atypical reactions to drugs such as barbiturates and halothane have already been found in individuals with certain genetic disorders and in people with certain normal genetic variations. Genetic variations can offer protection as well as susceptibility to industrial chemicals. One hypothesis is that the workers' genetic makeups account for the fact that under the same working conditions some workers develop a certain disease while others do not. This kind of knowledge can lead to the prevention of reproductive damage, but such practices raise serious ethical and social questions.

CHAPTER 2 APPENDIX 2: GENETIC TESTING IN THE WORKPLACE

Genetic testing in the workplace has two facets – genetic monitoring and genetic screening. Genetic monitoring involves examining individuals periodically for environmentally induced changes in their genetic material. Genetic screening tests individuals for certain inherited traits. The assumption is that either genetic predisposition, exposure to environmental agents, or a combination of both may predispose the worker to occupational diseases. Changes in the egg or sperm could then cause a baby to be born with birth defects.

Genetic monitoring involves collecting blood or body fluids from a group of workers periodically to assess whether genetic damage of the cell has occurred. The procedure focuses on the risk for the exposed group as a whole, because scientists cannot tell which individuals in the group face an increased risk. Ideally, genetic monitoring, if sufficiently improved and used to benefit the worker, could act as an early warning system by indicating that exposures to suspected reproductive or other health hazards are too high or that a previously unsuspected substance is causing harm.

In addition to specific genetic monitoring, there are four other kinds of monitoring activities relevant to detecting reproductive hazards in the workplace:

1. **Environmental monitoring** (EM) – attempts to provide a quantitative estimate of the dose of toxic exposure. It measures concentrations of the agent and not the amount inhaled or ingested by the exposed worker. This is referred to in the technical literature as a measure of 'intake'.
2. **Biological monitoring** (BM) – attempts to measure the actual total 'uptake' (intake × absorption) of the toxic agent by the exposed worker. This could occur through several pathways simultaneously (mouth, skin, lungs).
3. **Health surveillance** (HS) – periodic medical, physiological and biochemical examination of exposed workers with the aim of preventing occupationally related diseases.
4. **Biological effect monitoring** (BEM) – attempts to measure and assess early biological effects of workplace substances whose relationship to harmful reproductive and other health impairment has not yet been firmly established.

3

The Workplace: How Hazardous Is It?

> My job as a telephone equipment installer involved pole-climbing, ladder-handling, and the running of 25 and 50 length cables.
>
> My doctor didn't realise what I was doing, but once he said, 'Whatever you are doing keep it up, you're in excellent physical condition.' The only time he did find out about my job was when I was going into my eighth month and I asked him if he would write me a note saying that I wasn't allowed to climb poles any more.
>
> Belinda, telephone equipment installer

While we may not all be as physically fit or as adventurous as Belinda, pregnancy is certainly no longer viewed as an unemployable condition. As women have carved out niches for themselves in every type of occupation, they have increasingly enquired whether their workplaces are conducive to having a healthy baby, especially after they have seen stories about clusters of miscarriages and birth defects among women in various industries.

Women who work in healthcare, industry, agriculture and the service sector can be at risk of reproductive harm. They come into contact with many different substances and no publicly available toxicity information exists for most chemicals. Control of Substances Hazardous to Health (COSHH) regulations provide a little protection as they state that no worker should be exposed to a toxic substance exceeding the prescribed occupational maximum exposure limits (MELs), but unfortunately, MELs have only been issued for a small fraction of the more than 100,000 chemicals in daily use. This doesn't mean that exposure to substances not yet tested is harmful, but there is no way of knowing which ones are harmful and how much exposure exposes you to danger. Therefore, pregnant workers, for the most part, have no way of knowing the health implications of the chemicals they work with, let alone the reproductive consequences. Too often only after a group of workers have become sick has the toxicity of the substance become known.

Aside from possible exposure to toxins, just about all work settings have problems with at least one aspect of the environment that, if not actually hazardous, certainly results in substantial discomfort – excessive noise, poor lighting, indoor air pollution, poorly designed workspaces and equipment, heavy lifting, extreme temperatures, unsafe and unclean facilities and, most importantly, physical, psychological, and social stress.

Asthma, chronic bronchitis and emphysema are increasing rapidly and nearly 30 per cent of the cases may be attributed to exposures on the job. Many women do not realise that pre-existing asthma may be made worse by workplace pollutants. Even general dusty conditions can be an irritant, particularly for women who are vulnerable to respiratory hazards. If you smoke cigarettes you are facing increased harm as the combination of cigarette smoke and workplace pollutants makes you sicker than if you are exposed to each separately.

These problems are not confined to heavy industry or factory work. The office worker who lifts a ream of computer paper from a bottom shelf or stands on a shaky chair in order to reach a box on a top shelf is at risk of a back injury. A receptionist who answers the phone at a car body shop is subjected to the same toxic fumes as the painter and welder. Pregnant workers, however, because of physiological changes due to pregnancy, temporarily become even more susceptible to potentially hazardous conditions faced by workers across a wide range of occupations.

Canadian researchers studying pregnant workers in 42 occupational groups found increased rates of miscarriages across all occupations for women whose jobs required heavy lifting, other strenuous physical effort and long hours of work. Exposure to noise, cold and extended periods of standing were also associated with a higher than expected miscarriage rate, and stress – physical, psychological and social – was related to adverse affects on pregnancy.

When you are pregnant you are likely to be sensitive to conditions that you formerly could take in your stride, and there are many UK regulations aimed at ensuring the health and safety of employees

Table 3.1 Examples of occupational conditions that may affect health during pregnancy

Problems	*Workplace Factors*	*Job Types*
Backache	Standing Lifting	Factory worker Supermarket shelf filler Waitress Teacher Nurse/hospital doctor or worker Hairdresser Agricultural/farm worker
Expanding size	Use of protective clothing Confined spaces	Food processing worker Cleaner Nurse/hospital doctor or worker Supermarket cashier
Frequency of urination	Difficulty in leaving work station	Telephonist Conveyer belt worker Nurse/hospital worker Driver Teacher
Morning sickness	Early shift work Nauseous smells	Cleaner Factory worker Food processing worker Nurse/doctor/hospital worker Agricultural/farm worker
Possible miscarriage	Infectious agents Chemical toxins	Laboratory worker Nurse/hospital doctor or worker Manufacturing worker Teacher Agricultural/farm worker
Tiredness	Overtime Evening work	Waitress Nurse/hospital doctor or worker Teacher Shop worker Office worker
Varicose veins/haemorrhoids	Standing/sitting Working in hot environments	Factory worker Teacher Shop worker Waitress Hairdresser Dry cleaner Office worker
Additional problems	Homeworking	User of machines and chemicals in unsafe conditions and for long hours
	Exposure to ionising radiation	Radiographer Nurse/hospital doctor or worker Dentist/vet
	Exposure to lead	Lead battery worker Artist Metal reclamation worker Demolition worker Potter

whether or not they are pregnant which may be particularly appropriate to you at this time. For example, some evidence from animal studies (see the section on noise and vibration later in this chapter) suggests that excessive noise may be harmful to your pregnancy; the Noise at Work Regulations 1989 protect all workers from excessive noise. Even if the level of noise turns out not to be officially excessive, you certainly will be a happier pregnant worker without it. In Europe there is a proposal for a Physical Agents at Work Directive covering noise, vibration and non-ionising electromagnetic radiation, which could lead to new UK regulations that supersede the existing requirements.

Electromagnetic Fields (EMFs)

Electromagnetic fields (EMFs) have received a great deal of media attention, but just what are they? They are invisible lines of force created whenever electricity is generated or used. They are produced by power lines, electric wiring, and electric equipment and appliances. Workers and community members are exposed to both electric and magnetic fields, but scientists are more concerned about exposure to magnetic fields. Some studies have shown increased leukaemia and cancer rates among workers exposed to high magnetic fields, but on the current available evidence scientists disagree about the harmful health effects of EMFs; the one thing that they do agree on is that more investigation is needed. As an interim measure, until we know more about whether they are harmful, employers and workers can try some simple and inexpensive measures to reduce exposure to EMFs:

- magnetic fields often drop off dramatically within about three feet of the source, so work stations can be moved out of the three feet range of strong EMF sources
- exposure times to EMFs can be reduced
- low-EMF designs for the layout of office power supplies can be installed.

While the probability of harm to you and your foetus is slight according to current knowledge, it would be prudent to stay out of strong magnetic fields as much as possible, especially when you are pregnant (see more about electromagnetic fields later in this chapter in the section on VDUs).

The COSHH regulations state the general legal framework for both controlling hazardous substances and protecting individuals exposed to them. Substances regulated by COSHH include:

- substances labelled as dangerous (very toxic, toxic, harmful, irritant or corrosive) under other statutory requirements
- agricultural pesticides and other chemicals used on farms
- substances with occupational exposure limits
- harmful micro-organisms
- virtually any material, mixture or compound arising from a work activity that is harmful to health.

New COSHH regulations were passed in 1994. These incorporate all previous COSHH legislation in a single legislative package. The new regulations also mark the implementation of the European Union (EU) Biological Agents Directive which expands existing COSHH provisions regarding control of harmful micro-organisms and sets new or revised Maximum Exposure Limits (MELs) for nine substance groups.

UK membership of the EU has been a boon to British workers. Even though many of the items were already covered by UK legislation, some new protective measures, such as those connected with the use of video display equipment, were accelerated by the European accords. The Management of Health and Safety at Work Regulations (MHSW) 1992 implement the European Health and Safety 'Framework' Directive and part of the Pregnant Workers Directive; these regulations give specific protection from reproductive hazards to pregnant workers. Other regulations implement EU health and safety directives covering workplace standards, work equipment, protective clothing, visual display unit (VDU) work and manual handling. (You can find further information about directives and UK regulations in the appendices at the end of the book.) In theory at least, you have much more protection at work if you are pregnant now than you would have had ten years ago.

As we all know, gaining legal rights is only a first step. Actually being able to use these rights when employers wish to interpret them loosely or evade them completely is another story. A 1993 Health and Safety Commission (HSC) survey of British firms indicated that the vast majority of companies at least attempted to meet the requirements set out in the new regulations, but a small minority remained noncompliant either wilfully or due to ignorance.

WHAT SCIENTISTS KNOW AND DON'T KNOW

The regulations cover many known hazards, but much is still unknown about the impact on human bodies of many of the substances and agents present in workplaces. While scientists do not have answers, they do have suspicions about many substances and agents based on animal and bacterial studies as well as the more meagre human evidence. The United States Congress Office of Technology Assessment (OTA) reviewed the research on many agents and substances which showed some indication of reproductive harm. The findings regarding human reproductive damage were inconclusive for almost all the agents and substances. There is general agreement on lead, ethylene oxide (ETO) used in sterilisers, fumigants and industrial processes, dibromochloropropane (DBCP) used in certain pesticides and ionising radiation as being definite reproductive workplace hazards, though there is also growing consensus about organic solvents and heavy metals.

PROTECTING YOURSELF AGAINST HARM

The primary emphasis up to now has been to protect the pregnant worker, sometimes by removing her from a hazardous workplace. But this type of policy is inadequate as reproductive hazards affecting genes and fertility can occur in either sex. Protecting reproduction includes protecting men as well and has to start before a couple attempts to conceive a child. The Trades Union Congress (TUC) recommends union contracts that give both men and women seeking to become biological parents the right to 'safe' jobs, and in its handbook *Women's Health at Risk* (TUC 1991) it lists key points at which reproductive harm can occur and a list of primary sources of possible work related reproductive harm:

Points at Which Reproductive Harm Can Occur

- Disruption of menstruation and ovulation
- Decrease in sex drive in male or female
- Genetic damage to either sperm or egg
- Problems of foetus implantation

- Miscarriage
- Abnormal foetal development
- Problems in childbirth
- Polluted breastmilk

Sources of Possible Work-Related Reproductive Harm

- Chemicals (mutagens, teratogens, foetotoxins and carcinogens)
- Cold stress
- Heat stress
- Lifting and carrying
- Micro-organisms (e.g., rubella, cytomegalovirus)
- Radiation (ionising, electromagnetic)
- Raised atmospheric and hydrostatic pressure
- Vibration (low frequency)
- Work stress (shiftwork, nightwork, long hours, excessive demands, poor interpersonal relations)

More information is given in the two tables in Appendix B at the end of the book. One lists the agents and substances reviewed by the Office of Technological Assessment (OTA), United States Congress and the second lists substances, their types of possible reproductive effects and whether the data come from human or animal studies.

If you think you are exposed to any of these substances or agents and if you are pregnant or considering getting pregnant ask your union safety representative for the latest information. Safety representatives have the right to study COSHH risk assessment documents as well as substance and product data sheets to see if the substance is considered to be a reproductive hazard.

If you do not belong to a union:

- contact the Health and Safety Executive or your local Environmental Health Officer (EHO); if the EHO fails to take adequate action, report the matter to your local councillor
- check with your general practitioner or midwife
- call the nearest university department of occupational or environmental medicine
- contact City Centre and the London Hazards Centre (addresses are in the Useful Addresses section at the end of the book).

Because carcinogens (substances and agents that cause cancer) and mutagens (substances and agents that cause genetic defects) are highly related, you may be helping to protect yourself against mutagenic exposure by protecting yourself against cancer causing risks. This, however, will not necessarily protect you against exposure to substances that may cause developmental defects (teratogens).

In no case should you be exposed to toxic substances exceeding the prescribed occupational exposure limits (OELs) under COSHH. The catch is, however, that while there are over 100,000 substances used in workplaces, OELs, as well as the MELs discussed previously, have been set for only a small proportion of these. The regulations implementing the EU Health and Safety Directives can provide much improved protection for British workers if they are enforced, but this is a big 'if' as it is not at all clear whether the monitoring of the various measures covered by the directives will be fully implemented. Instead, the narrowest interpretation might be used.

Workers who plan to become pregnant desire precise information about potential workplace risks and hazards and it is frustrating that this is not yet known – this uncertainty can make it difficult to know what to do. However, women are learning how to make judgements and decisions based on the best available information. This chapter provides information on workplace conditions and substances that are of concern to working women who want to make their pregnancies as comfortable and healthy as possible. Hopefully, in the future more of you will be able to obtain this information from your primary healthcare providers; a research report was published by the HSE in 1994 on the need for training to enable practice nurses to offer occupational health information advice to clients at health promotion clinics.

A key point to remember is that if the evidence that you find is very weak, do not worry unduly about the exposure, despite the fact that so often exposures deemed to be harmless turn out to be harmful later. Constant worry can cause stress which is already known to be harmful to your health and which may be potentially more damaging than the slight possibility of harm caused by exposure to a given agent. *Minimise your risks as much as possible, but remember that most women do give birth to healthy children.*

Specific Work Places

AGRICULTURAL WORKERS

(Also see the later section on pesticides used in the home.)

In the mid-1980s approximately 84,000 people worked as agricultural and farm workers in Great Britain, including young people and pregnant women. Many of them are regularly exposed to pesticides. This is an extremely serious problem as pesticides once thought of as saviours are now considered possible carcinogens and reproductive hazards. Pesticides and fumigation are covered by the Control of Pesticides Regulations 1986 (COPR) under the Food and Environment Protection Act 1985.

COSSH regulations cover the safe use of pesticides in all types of industry, including agriculture and horticulture, and the TUC has argued that COSHH takes precedence over the Food and Environmental Protection Act. Under COSHH, compliance in controlling hazardous exposure involves the substitution of substances and/or engineering controls, while the Food and Environment Protection Act only requires distribution of protective clothing and respiratory equipment. But even within each of these regulations, disagreements about risks and how they should be handled occur.

In 1991, for example, British government scientists found pesticide residues in carrots at ten times the maximum permitted levels due to changes in farming practices which increased the spraying of the carrot crop. Instead of insisting that farmers drop the practices that caused the high levels, the Working Party on Pesticide Residues decided instead to raise the permitted level of triazophos pesticide claiming it remained well within the safety limit.

In the United States a Department of Health and Human Services study found that one of every three farm workers in southern New Jersey claimed to have been accidentally sprayed by an aeroplane or tractor rig with unsafe levels of chemical pesticides. This was in just one small section of the country and there is no reason to believe that it is atypical, meaning that large numbers of pregnant workers are likely to be exposed to big accidental doses and lower chronic doses of toxic pesticides.

The side-effects of such pesticide exposure or poisoning are usually severe and immediate and take the form of nausea, occasional vomiting, dizziness, chest pains, eye problems, skin rashes and flu-like symptoms. If you are pregnant and work or live in areas where pesticides are heavily used and feel any of the above symptoms, inform your GP or midwife immediately.

Most cases of acute or chronic long exposure pesticide poisonings are not even diagnosed as such. Doctors seldom ask about occupational exposures and the same symptoms can result from a wide variety of other causes. Some mimic normal pregnancy symptoms and so are not attributed to the toxic effects of pesticides. Even if such toxic effects are suspected, the worker and farmer often do not know the names of the chemicals to which they are regularly exposed.

If you live in an agricultural community, even if you are not a farm worker or pregnant, you should be alert to the possibility of pesticide exposure as aerial spraying drifts with the wind, contaminating vegetables grown in back gardens and residents in the surrounding area. If you are pregnant, it may be wise to find out the dates for massive aerial spraying and either leave the area for the day or stay indoors with windows shut.

> I didn't really think about it at the time, but I lived where there was a lot of crop dusting. Every winter I'd get a sore throat as a reaction to this cotton defoliant they were spraying around the middle of December. That had nothing to do with the job, it had more to do with where I lived. When I remembered, I would always ask my doctor and he was reassuring even when I was pregnant and due to deliver in January.
>
> Kitty, US teacher

Pesticides are used in non-agricultural settings as well, particularly in jobs which involve work with wood. Unions in the UK have even managed to negotiate bans on the use of deadly pesticides. For example, carpenters and other labourers working for the Hackney Council on building projects negotiated a ban on five chemicals used for wood preservation: dieldrin, pentachlorophenol (PCP), tributyl tin oxide (TBTO), copper chrome arsenate (CCA) and gamma HCH (lindane) plus the weed killer paraquat. The safety representatives found these chemicals to be central nervous system poisons, reproductive hazards, skin and respiratory irritants and cancer agents.

The Chemical (Hazard Information and Packaging for Supply) Regulations 1994 require manufacturers and suppliers to provide data sheets on their chemicals to users – this means employers and workers. Enough information about chemicals classified as dangerous (and this includes almost all pesticides) is to be provided to enable users to take proper precautions. If you are a member of the community outside of the workplace and believe that you have been exposed, or if you do not even know what pesticide you might have been exposed to, you will probably have to take a more indirect route to find the information you need. Enlisting the aid of your GP or contacting your Citizens Advice Bureau may be good first steps.

HEALTHCARE JOBS

> I was a foetal medicine midwife which involved follow ups of women with problem pregnancies. I was not involved in the delivery of babies. My job required a lot of walking around and standing, but no lifting. During the first three months I was very nauseous so I would take a few sick days. I had to eat frequently so I brought biscuits to work. The staff was extremely supportive.
>
> Bridget, foetal medicine midwife

The job of a healthcare worker is to heal. It is ironic that the process of healing can cause harm, particularly for those working in a hospital or laboratory. If this applies to you, be especially careful to protect yourself as much as possible from coming into contact with infectious material or exposing yourself to X-rays or other toxins as some may cause mutagenic, teratogenic and carcinogenic changes, illnesses, discomfort and injury. (Unfortunately, using latex gloves, introduced as a protective measure, is now causing harm to as many as 10 per cent of healthcare workers. Those sensitive to the latex can develop what is called contact dermatitis which can develop into chronic skin disease. In addition, a small number of workers have developed latex-related asthma, an even more worrying problem.)

> I take care of people in the geriatric and psychiatric sections. I bathe the people and put pampers on them in the geriatric ward, so you have to lift them a little. When I became pregnant, I was told not to lift patients and not to go near patients with infections. You have to be careful with the psychiatric patients. They can hit you. I always was careful and never got hit.
>
> Zena, US auxiliary nurse

Healthcare is also a high stress occupation. Curing and caring provide sustenance to the professional, but illness and death are anxiety-producing. If you add disagreements among patients, families, doctors and nurses, the atmosphere can turn into a pressure cooker, particularly in intensive care units and emergency rooms. Workers become exhausted and can develop headaches, insomnia, ulcers, gastrointestinal problems, high blood pressure, and heart disease – conditions bad for anyone, but even worse for a pregnant woman and her unborn child.

Healthcare institutions tend to neglect the health of their employees in their attempt to improve the health of their patients; not a very sensible approach. What they save in money, they lose in high staff turnover and burn-out. Luckily the public services union UNISON is very active, emphasising education, training and prevention as well as the usual wage and conditions of employment issues.

The recent changes in the National Health Service have added additional levels of bureaucracy including increased paperwork. While some of this new accountability and evaluation of performance was badly needed, the rapid shifts in power between medical and administrative and accounting staffs have provoked much friction and stress. This has been particularly true at some of the first NHS Trust hospitals.

> In some respects, the NHS is one of the worst places to work. It's under such a strain that it is very difficult at the moment to allow for anyone pregnant or anyone who has a problem. I'm very lucky in the department I work in because the people are very supportive. They have been flexible and supportive while the system as a whole doesn't need to be. So many of the checks and accountability seem to be for the sake of it. They sounded good on paper but are not very beneficial. They don't seem to achieve anything.
>
> Heather, hospital clinic administrator

Injuries, radiation, high energy exposures, infections and chemical hazards are fairly common in hospitals. They are more serious if you are pregnant and to be avoided as much as possible. If you think you are particularly vulnerable to any of them, see the further information in the appendix to this chapter and contact your union safety representative or the health and safety office of UNISON. Some particular healthcare jobs are more hazardous than others.

The Cancer Treatment Unit

Until recently, we did not realise the health implications of dispensing cytotoxic anti-cancer drugs. They are now linked to the development of cancer in the health provider and birth defects in her unborn child. Swedish researchers reported finding increased levels of chromosome abnormalities associated with long-term handling of chemotherapeutic agents. Prior research found that chromosome damage was reversed after measures to limit exposure were installed. A Finnish study found that exposure to cytotoxic drugs may cause liver damage. Moreover, it is a type of damage not usually picked up by ordinary liver function tests until it is severe. Anecdotal reports by nurses also describe milder effects – facial flushing, light-headedness and dizziness.

There are guidelines for the safe handling of highly toxic chemotherapy agents which also cover the preparation, administration and waste disposal of cytotoxic anti-cancer drugs. The best advice, especially if you are pregnant, is to be extremely careful and to wear face masks and gloves when handling any chemotherapeutic agents. One word of caution, however. A report in the American Journal of Hospital Pharmacy warned that commonly used material in surgical gloves might be permeable to anti-cancer drugs. The researchers found that the one commonly used chemotherapeutic agent they tested permeated the glove within five minutes and the amount increased over time. Do not just assume that because you are wearing gloves you are completely safe. You may have to change to a surgical glove made of different material or 'double glove' – wear two pairs.

The Dental Surgery

The dental surgery is a microcosm of the hospital setting, with potential exposure to a combination of chemical, physical and biological agents. A large proportion of dentists practising in the UK are women, many of whom are in their childbearing years, and dental assistants and hygienists are also mainly women. These personnel may be exposed to primary herpes and hepatitis as well as other micro-organisms present in the patient's saliva. Waste anaesthetic gases, airborne mineral dusts resulting from high speed grinding of dental material, ethylene oxide from sterilisers, mercury, ionising radiation, high noise levels from drills, backaches from standing in awkward positions are among the more common hazards.

> When I took X-rays I was afraid of the scatter radiation. We tried to work it out that the other girl would take more of the X-rays. Sometimes it worked. Sometimes it didn't.
>
> Rory, dental assistant

The following precautions should be taken if you work in a dental surgery. Most of the suggestions for improved safety for X-ray and lab workers presented in the following sections also apply to dental personnel.

- Obtain a complete patient health history and update it at each visit.
- Use a rubber dam to limit the spread of aerosolised saliva.
- Wear a face mask and surgical gloves while working on a patient.
- Buy handpieces and air–water syringes that can be heat sterilised.
- Have your patient routinely rinse his/her mouth prior to the start of the dental procedure.
- Use only X-ray film rated at least speed group d.
- Check whether the equipment in your office is periodically inspected.

The Laboratory

Women of childbearing age make up the majority of clinical and research laboratory workers. Laboratories have many potential work hazards. The Pasteur Institute in Paris, for example, found an increased risk of cancer, especially brain tumours, among their biomedical laboratory workers under the age of 50. In England, one male and two female technicians from the same laboratory at a district hospital developed brain tumours within a two-year period. Although this is an unusually high number of brain tumours so many factors, such as genetic susceptibility or even mere coincidence, can be responsible that it will take very sophisticated epidemiological studies to determine whether, in fact, there is a genuine occupational link. The International Agency for Research on Cancer has initiated a study to examine the cause of death of laboratory workers, with the British part funded by the Medical Research Council and the Health and Safety Executive.

Some laboratory associated illnesses come directly from patients' infections such as hepatitis-B (discussed later in this section) and some from the materials and techniques used to determine the illnesses. For example, mercury, lead and carbonate compounds used in laboratory procedures are known to cause birth defects in animals and humans. Dioxane, a dehydrating agent used for slide preparations, can cause liver and kidney damage. Radioisotopes and radioactive patient specimens can also be harmful to the unborn child.

Most of these risks can be prevented by careful work habits combined with stringent enforcement of safety regulations and standards regarding the maintenance of both the lab equipment and the laboratory itself. It is easy for routine to allow you to be a little careless, but whereas this can ease the stress of a monotonous job a little extra effort can also prevent serious harm to you and your baby.

> In my job I take a risk anyway. We work with blood. It's not always in the bag. You never know when the bag leaks. They were having the AIDS scare while I was pregnant. It was touchy for a while and hepatitis is a possibility, too. I'm sure it was a greater risk being pregnant. There was nothing I could do about these risks. I could quit early.
>
> Maria, US laboratory technician

UNISON advises that all clinical laboratories draw up a local code of practice based on the Health and Safety Executive's 'Code of Practice for the Prevention of Infection in Clinical Laboratories and Post-Mortem Rooms'.

How Diseases are Transferred in the Laboratory

- Handling, analysing and disposing of biological material, e.g. blood, tissue.
- Inadequate labelling and packaging of biological material.
- Uncovered or overfilled rubbish bins containing carelessly discarded contaminated specimens.
- Accidental pricking of skin with contaminated instruments.
- Direct exposure to contaminated blood.

What To Do

- Use gloves when handling infectious specimens.
- Do not pipette by mouth. Use disposable pipettes or pipette tips.
- Wear protective clothing, e.g. lab coats.
- Use effective ventilation hooding.

- Do not eat, drink or smoke in the lab area. Remove lab coat prior to eating, drinking, or smoking in areas designated for those purposes.
- Clean up all spills quickly and thoroughly with sterilising solution.
- Insist that COSHH regulations are available and enforced. Make sure that your employer provides a complete list of all chemicals used and their toxic effects.

Laundry, Housekeeping, Maintenance and Catering Services

(For further information see the section on dry cleaning and laundry workers in the service sector.)

If you are an ancillary staff member in the hospital – porter, domestic, laundry or catering worker – you may not have been adequately trained to avoid contamination on your job. This is particularly true if you have been hired by a private contractor. If you are pregnant, therefore, you may be even more vulnerable to harmful health and reproductive effects than the professional staff who have usually had some education about the risks and hazards involved in their work.

If you are a porter or collect dirty laundry, be careful to avoid contact with infectious waste material which is improperly discarded and not labelled as being infectious. Make sure that you never touch contaminated laundry with your bare hands.

Sometimes, tired or poorly trained workers get careless and just throw infectious materials, instruments, body fluids and tissues into waste baskets or roll them up in laundry. Ancillary staff need to inquire whether patients are infectious. If they are, the cleaning staff must continue with the strict cleaning procedures designed for infectious areas even after the patients are discharged. UNISON can help you by ensuring that the hospital provides education and training and maintains strict health and safety practices.

The Operating Room

Several research studies in Europe and the United States have linked exposure to waste anaesthetic gases in the operating room to an increase in spontaneous abortions and birth defects among nurses, physicians and wives of physicians. A nationwide study by the

American Society of Anesthesiologists also found an increased risk of cancer, kidney and liver disease in women working in the operating room. The same connection was found among dental personnel who used gas anaesthetics in oral surgery and rehabilitative dentistry. However, other scientists who reviewed the major studies found that significant flaws in study design and investigation procedure made it difficult to conclude that occupational exposure to anaesthetic gases causes increases in miscarriages and birth defects, and a controversy about the health effects of exposure to anaesthetic gases continues.

> I had two miscarriages before my first baby. I was working in the operating room at the time, where I was exposed to anaesthetic gases. Some of the anaesthesiologists had anaesthesia machines that were kind of antiquated and there was no way of venting the gases out of the room or they refused to do it because they thought there was no problem. I was really angry at them. They weren't the ones who had the miscarriages and even if it didn't cause a problem, it really wasn't a big deal for them to hook their machines up to the suction equipment.
>
> The hospital did not have any policy and nobody would stand behind me. The nursing director wouldn't and it was easier for me to leave than to get someone to back me. It was easier for me to just transfer out. It is kind of ridiculous that you have to go to that extreme. I was under a lot of stress in terms of what my environment was causing. I was really upset.
>
> Clarissa, US operating room nurse

In response to the initial findings, many hospitals improved the scavenger systems to remove most of the anaesthesia that had leaked into the air. Not all newly installed equipment work equally well, however; you should ask your union safety representative to check whether your employer regularly monitors for accidental gas leaks.

X-ray equipment is also often used in operating rooms, and pregnant operating room personnel could be exposed to escaping radiation as well as escaping anaesthetic vapour. Clarissa was one nurse who may have been exposed to both.

> They did a lot of procedures that used an X-ray machine in the operating room throughout the whole procedure. All the circulating nurses and the scrub nurses wore lead aprons for protection. But

you know, there is still some radiation. The X-ray techs that came in to do the X-rays were always monitored. They wore little badges, but the nurses and doctors were never monitored.

They did not allow people that were pregnant to go in those rooms. However, the two pregnancies that I had were not planned, so how many procedures I did before I even knew I was pregnant, I don't know.

Clarissa, US operating room nurse

The Sterilisation Department

Sterilisation of hospital facilities and equipment is given high priority. Now high priority has to be given to the sterilisers themselves. Ethylene oxide (EtO) is one of the main sterilising agents used in the hospital. In laboratory studies EtO has both affected the reproductive capacity of male mice and increased mutations of genetic material in hamster cells. Furthermore, workers in Sweden who were exposed to only low levels of EtO in the process of producing the chemical were found to suffer from leukaemia and stomach cancer ten times as frequently as the national Swedish rates. Even though the use of EtO in hospitals represents only a small percentage of the total production of EtO, healthcare uses present the major source of human exposure to this chemical.

I worked in the sterilising department of the hospital. Ethylene oxide was used for sterilising instruments, sheets and blankets in the steam steriliser. It is a poisonous gas. We wore masks. We also wore buttons to see if gas had escaped. These buttons show how much gas you are exposed to. Gas never escaped while I worked there. The job was physical in the sense that large racks had to be pushed into the room steriliser and they were extremely heavy.

Pam, US steriliser of medical instruments

In sterilising material and equipment, workers are exposed to relatively high levels of EtO for brief periods. This usually occurs during the transfer of material from the EtO steriliser to the aerator unit. Danger of reproductive injury for this short period – high dose contamination typical of the hospital situation is a problem. Many cases of miscarriages have been reported among pregnant women exposed to periodic short-term, high doses of EtO while working with sterilisation equipment. It is obvious that a standard for an average level of exposure does not protect you under these conditions. If you are trying to conceive or are pregnant, ask for a transfer to another job if at all possible.

X-ray and Nuclear Medicine Department

Those of you who work with X-rays or in the nuclear medicine department need to be concerned about being accidentally exposed to ionising radiation. Ionising radiation is energy that is transmitted in wave or particle form and is capable of causing ionisation of atoms or molecules in the irradiated tissue. It is known to exert strong effects on the developing embryo, foetus and child as well as affecting normal reproductive functioning in men and women. High doses of ionising radiation, much higher than doses likely to be found among employees in the medical field, impair testicular function in males. There is also some indirect evidence that it is associated with lowered sex drive and less healthy sperm. So if you are planning to have a baby and your husband or partner works in one of these areas, he also needs to be careful to avoid reproductive harm.

Female offspring exposed to ionising radiation while in the womb can suffer from reproductive disorders years later. These may be abnormalities in their endocrine systems ultimately leading to infertility or abnormal development during puberty. High doses can cause sterility and can initiate menopause. These effects are more common to patients receiving treatment than to professionals treating them. Nevertheless, it is known that if you are exposed to levels of greater than 20 rads while you are pregnant, your baby may be born with birth defects. Lower exposure from 1–10 rads is associated with increased risk of mental retardation, leukaemia and cancer in your offspring. We do not know what effect, if any, extremely low doses of ionising radiation have on reproduction. Recent re-evaluation of data on radiation exposure of victims of the Hiroshima and Nagasaki atomic bombs indicate that the maximum allowable worker exposure should be reduced.

Most of the recent controversy about the health and reproductive repercussions of exposure to radiation has arisen from investigation of the British nuclear industry safety standards, particularly regarding the statistical association between leukaemia in children of Sellafield workers and radiation doses received by their fathers prior to conception (see Chapter 2 for additional information).

While the Sellafield study is aimed at workers and their families in the nuclear industry, medical personnel also should remember to take proper precautions at all times and X-ray technicians need to be shielded by lead. Make sure that the machinery you work with is maintained properly and that there are no leaks. Portable X-ray machines have the greatest likelihood of leaking and extra-special precautions should be taken when working with them. This advice holds if you are working in a hospital, laboratory or doctor's or dentist's surgery, but precautions should also be taken if you work with radioactive isotopes or other diagnostic or therapeutic radioactive material. At present the use of lasers, a form of non-ionising radiation used as diagnostic and surgical tools, has not been shown to be harmful to reproductive functioning.

> The times to be extra careful to avoid exposure is when you are attempting to become pregnant or very early in your pregnancy. The main thing that really scared me was to walk through the radiology department to go to lunch because they have to keep the doors open. I never tried to find out about any of the risks. I would take a different route so that I avoided radiology completely.
>
> Jenny, US medical technologist

Infections

(Also see sections on teachers, saleswomen and flight attendants for more information.)

Workers in all areas of the healthcare field are also exposed to infections as part of their jobs, some of which can be very damaging and life threatening. The most important of the infections to guard against are HIV, hepatitis-B and, in recent years, tuberculosis and hepatitis-C.

HIV and AIDS

AIDS is a disease that poses a potential hazard to healthcare personnel and is of particular concern to pregnant workers as an HIV-positive woman can pass the virus on to her unborn child. However, the HIV (AIDS) virus is only spread through blood and sexual fluids rather than casual contact, and so far only a handful of healthcare workers not definitely known to belong to a high risk group (mainly intravenous drug users, homosexuals and haemophiliacs) have developed AIDS. We have been told so often in the past that our worries about potential harm from occupational or environmental hazards were misplaced only to learn at a later date that our concerns were justified after all, that we have become wary of believing these reassurances. In the case of AIDS, however, careful investigation of close, non-sexual family contacts over long periods of time have shown no cases of transmission.

While the risk of contracting the HIV virus from ordinary patient contact is minuscule, it is crucial for you to follow the designated procedures set up to protect healthcare workers from exposure to such a deadly disease:

- a strict adherence to hospital infection control guidelines regarding sterilisation, housekeeping and disposal of infectious waste
- the use of personal protective equipment such as gloves, gowns, masks and eye coverings wherever contact with blood and body fluids is expected; if body contact occurs by accident, the worker should immediately and thoroughly wash the area
- the use of convenient puncture-resistant containers for disposal of sharp items and careful handling of sharp items
- the ready availability of mouth pieces, resuscitation bags or other devices to avoid mouth-to-mouth resuscitation.

Only staff trained in good care standards, disinfection procedures and protection precautions should care for AIDS patients. If you feel ill, do not care for AIDS or HIV patients. These patients have lower levels of immunity and instead of helping them, your care can put them at risk of catching an infection. It is also unwise for pregnant staff members to work with AIDS patients because some of these patients may excrete high levels of cytomegalovirus which could cause birth defects.

Guides have been written explaining the procedures healthcare personnel should follow in order to avoid accidental exposure to infected blood or other body fluids. When you use any AIDS prevention guide, check the date of original publication or revision in order to obtain up-to-date information. Material on AIDS in employment can be obtained from the Trades Union Congress (TUC, Congress House, Great Russell Street, London WC1B 3LS); the Health Education Authority (Hamilton House, Mabledon Place, London WC1H 9TX);

and UNISON, which has an AIDS guideline booklet which provides detailed advice: *HIV and AIDS – A Guide for Branches* (stock number 866 from UNISON, Civic House, 20 Grand Depot Road, London SE18 6SF).

Hepatitis-B

Hepatitis-B is caused by a virus and causes an inflammatory condition of the liver. Ninety-five per cent of people who contract hepatitis-B recover completely, but for a small percentage the disease is life threatening. Hepatitis-B can be a major risk for the health service staff, including physicians and nurses. Usually the disease is contracted through the accidental penetration of used needle sticks or other glass instruments or through infected blood. But one of the biggest problems faced by the health staff is that the disease can be passed on by asymptomatic patients.

There are two types of vaccine for hepatitis-B, the preventative kind for healthy workers who are at risk from their jobs and one for those who have already been contaminated by the virus. If your risk assessment has identified you as being at risk of exposure, you should be given the first type. Hepatitis-B has been recognised as a 'prescribed industrial disease' so that if you contracted the disease through work and are off sick, you can claim certain state benefits including the Industrial Injuries Disablement Benefit.

If you are contaminated by the hepatitis virus through your work on a health staff, you must immediately report the incident and seek the second type of vaccination treatment (you must obtain the hepatitis-B immunoglobin vaccine from the Public Health Laboratory Service within 24 hours of contact). In case of a puncture wound, you should try and make the wound bleed and then wash with soap and water. If there is a chance that you have been in contact with material contaminated with the hepatitis-B virus, wash the area of contact thoroughly.

Your workplace should offer the following precautions:

- all staff should understand the risk of hepatitis-B and be given adequate health and safety training (including private contractors' staff)
- regular workplace inspections should be undertaken by the health and safety representatives to check equipment, cleaning

and decontamination and to make sure that infection control measures are applied and carried out

- protective clothing should be available and worn when needed, and cuts covered.
- waste material should be properly disposed of, soiled linen correctly bagged and specimens for laboratory analysis placed in sterile containers with screw top lids, bagged and clearly labelled before transportation
- food and personal possessions should not be taken into areas with hepatitis-B risks.

HOMEWORKERS, PAID AND UNPAID

While we realise that we may face hazards in our workplace, we tend to think of our homes as sanctuaries. It comes as a shock to find out how wrong we are – home may be a haven, but it is increasingly a polluted one. Homemakers and household workers are exposed to air pollution and face insidious hazards such as noise from household appliances, injuries arising from moving and lifting heavy objects around the home, electric shocks from loose wires, radiation from leaking microwave ovens, tripping over toys the children have left out and stress from long hours and heavy demands of housework and childcare.

Household Hazards

- **Noise** – excessive noise from household appliances such as washing machines, dishwashers, clothes dryers and vacuum cleaners, may lead to hearing loss.
- **Poorly maintained equipment** – can cause electric shocks.
- **Repetitive tasks** – performing these in uncomfortable positions, and excessive standing and lifting heavy objects can cause muscle and bone injuries.
- **Long hours, low or no pay, working in isolation** – can lead to increase in stress and stress-related illnesses such as in cholesterol levels, hypertension, headaches, frustration and depression.
- **Microwave ovens** – continued exposure to leakage can be hazardous to eyes and reproductive system.

- **Formaldehyde** – a carcinogen in animals which may also enhance the effects of other carcinogens and mutagens is found in home products, cigarette smoke and indoor air pollution.
- **Drain cleaners** – can cause burns and can produce toxic gases if combined with bleach or toilet bowl cleaners.
- **Oven cleaners** – give off fumes that are irritating to breathe.
- **Ammonia and bleach** – can cause eye and lung irritation and can form toxic gas if mixed with each other.
- **Cleaning fluids** – may contain organic solvents which can cause irritations. Some of these substances can cause cancer.
- **Aerosol sprays** – can irritate lungs.
- **Waxes and polishes** – can irritate lungs and nasal passages.
- **Pesticides** – can cause serious poisoning.
- **Accidents** – burns from cooking, broken bones from slipping on wet floors and tripping over clutter.

Prevention

- Substitute less toxic products.
- Read the label.
- Keep appliances and electric circuits in good condition.
- Use gloves when handling potentially hazardous products.
- Clean up clutter around the house.

Source: Adapted from Centre for Science in the Public Interest, *The Household Pollutant Guide*. New York: Anchor Books, 1978.

Some women take care of their own families and homes and are not considered to be part of the paid work force. Others serve as household helps in other people's homes. Still others bring in outside work to do in their own homes. While all women who are part of the paid work force but work in their homes (homeworkers) are covered by the Health and Safety at Work Act and its regulations, there has only been a minimum effort by employers to comply with basic health and safety laws. Homeworkers and their families can therefore be exposed to the solvents and soldering fumes used in electronics and toy manufacture, while the homeworkers themselves are at risk of repetitive strain injury (RSI) from keyboard use, sewing and other types of piece work.

The HSE has produced a free booklet for homeworkers clarifying their rights, called *Homeworking: Guidance for Employers and Employees on Health and Safety*. (Unfortunately it is only available in English, although many homeworkers come from ethnic minorities and frequently read little English.)

Air Pollution in the Home

Indoor air pollution is frequently more severe than outdoor air pollution, and there are several reasons for the increased hazardousness of the air in our homes. These hazards mainly come from three sources.

- radon – an odourless, colourless gas which is a natural form of radiation resulting from a breakdown of uranium in soil, rocks and water
- changes in building and home furnishing materials – the substitution of plywood, particle board construction containing formaldehyde, and plastics for wood
- the growing use of chemicals in cosmetics, aerosol sprays, insecticides and cleaning products.

In the wake of the oil crisis in the early 1970s, soaring fuel costs provided the incentive for millions of owners of older homes to insulate their houses. Unfortunately, a 'tighter' home means restricted air circulation leading to similar problems as those found in office buildings. In older homes, all indoor air is exchanged with fresh outdoor air about once an hour. In extremely air-tight houses, complete replacement can take as long as ten hours, resulting in large build-ups of potentially harmful air-borne substances.

You can reduce indoor air pollution in your home by some simple measures – mainly involving the substitution of less toxic products. While reading labels is time consuming and a nuisance, taking the time to do so and so to find products with natural ingredients and fibres is one of the few things easily within everyone's power, and this information can help you have a healthy pregnancy. You will find that it is not always necessary to use commercial cleaners and detergents most of which should be kept out of reach of children, for example; it is remarkable what soap and water, baking powder or vinegar solutions can accomplish.

In addition, be wary of synthetics. Try to purchase wood rather than plywood or plastic furniture. If cost is a problem, buy second-hand wood furniture rather than new synthetic substitutes. Buy natural fabrics like cotton and wool. Many new fabrics and carpets are made from petroleum and other chemical bases which have been largely responsible for the black toxic smoke that has killed many people in recent hotel fires.

Be careful of carbon monoxide which can seep into the house through cracks in a wall when the motor of a car is kept running in a closed attached garage or from wood-burning fireplaces from which run partially clogged chimneys. Carbon monoxide is also one of the by-products of cigarette smoke, and if several smokers are present cigarette smoke can raise the level of carbon monoxide above the acceptable level even in a room which is adequately ventilated for most purposes. Carbon monoxide not only can cause nausea, dizziness and even death in the person breathing it in, but recent animal research has shown that it is also harmful to a foetus. It can cross the placenta and by doing so lowers the amount of oxygen that the foetus can obtain from its mother and hinders the ability of the haemoglobin to release the oxygen it carries.

Radon: Natural Radiation Seeps Into Your Home

One of the indoor air pollutants thought to be very harmful is radon gas, the natural form of radiation that is linked to lung cancer. Only in the last decade have scientists become concerned about the possible effects of radon build-up in private homes.

Originally, the radon problem was thought to be localised in areas where the soil contained radium or uranium, but it is now known that the problem is more geographically widespread as soil permeability may be a more important factor than the uranium content. The five-year British childhood leukaemia study described in Chapter 2 is including in its investigation the radon level of the homes of the children in the study. Researchers believe that if leukaemia patients lived in houses with higher levels of radon gas than did children in the control group (those who have not contracted leukaemia) then the connection between exposure to radiation and leukaemia cannot be just a coincidence.

As usual, experts disagree as to what is considered to be a high radon level. Radon is measured in picicuries per litre (pci/l). Many scientists suggest that radon levels in houses should be under 4.0 pci/l, while others believe that remedial efforts should be concentrated on homes with radon levels over the 10.0 pci/l level. If you are a smoker you should be even more concerned about the radon level in your home as some research indicates that any health risk from radon is increased by its combined effect with cigarette smoke – a 'synergistic' relationship. Women and children often spend more time in their homes than men and are, therefore, likely to be more at-risk from exposure to indoor air pollution. But don't panic. It is not clear how much of a hazard low levels of air pollutants are and the estimates of health effects of high levels are usually based on lifetime exposure. This is another area in which you can do something to help prevent danger.

Moderate radon levels can often be reduced easily and relatively inexpensively by sealing cracks and crevices in the basement, covering open sump holes or keeping some windows open. Higher levels will require more expensive and sophisticated techniques such as sub-slab ventilation techniques and air-to-air heat exchangers. To carry out this work, contact a reputable contractor who is experienced in radon remediation.

Household Cleaning Supplies: Pre-Detergent Possibilities

We worry a great deal about what might hurt our unborn children. A large weight is removed when we learn that our babies are born healthy. Yet we often pay too little attention to what might hurt our children after they are born. Most often, we are careless because we do not know that many common cleaning supplies can be toxic, especially to young children. An old-fashioned punishment was to wash out children's mouths with soap as a punishment for using curse words. They survived to tell the tales, though few would support such a punishment today. Now we wash with detergents and think of detergents as being as benign as soap. No such luck!

Detergents contain such chemicals as linear alkylate sulfonate and ethyl alcohol; dishwashing powders contain sodium metasilicate, which causes burns. Each year children are poisoned by drinking household products – more from dishwashing and washing machine detergents than from anything else. This could be because they are packaged in brightly coloured plastic bottles and boxes and left in easily reachable locations, a hazard which can easily be reduced by keeping cleaning supplies in locked closets or using six inexpensive and effective substances that our grandmothers used: soap, vinegar, baking soda, ammonia, washing soda and borax.

Pesticides: Are They Worse Than the Insects?

From the time they lived in houses, wore clothes and stored food, our ancestors had to contend with animals and insects wanting to share their quarters and possessions. They discovered many successful methods of getting rid of these pests which, while perhaps less efficient than the pesticides we now use, are also less toxic to us and are not likely to generate pesticide immune insects. Two of the most powerful 'anti-insect weapons' our grandparents used were cleanliness and sunlight. Now that so many of us hold two full-time jobs, one outside the home and one inside the home, we have had to skimp on chores such as the outdoor airing of bedding and thorough closet cleaning which were considered a must in our grandmothers' and great-grandmothers' days. The ritual of spring cleaning not only resulted in neater, cleaner, fresher smelling houses, it also eliminated pests who like to live in warm, humid, and dirty crevices, among minute biscuit crumbs and spilled milk stains and in old clothes not worn for years.

Bernice Lifton in her book *Bug Busters: Getting Rid of Household Pests Without Chemicals* (Lifton 1991) offers a seven point strategy: build them out; starve them out; clean them out; shake them up; trap them; barricade them; repel them.

Most of us abhor cockroaches, ants, wasps, mosquitoes, fleas, lice, bedbugs, rodents, waterbugs and spiders and want to get rid of them quickly and forever, but we often use inappropriate methods to accomplish this. A massive overkill assault with toxic commercial pesticides, particularly those in aerosol cans where pesticide vapour can float in the air and settle on you as well as the insects, is not the answer. The recommended non-toxic methods are slower, but it is much more important to insure your own and your unborn child's health than to rid your house in one swoop of the pests you have already been living with for a while.

Pesticides fall into five chemical categories: chlorinated hydrocarbons, organophosphates, carbamates, botanicals and inorganic compounds. All pesticides are somewhat toxic, but each chemical group contains some extremely toxic pesticides and some that are safer for humans, if used with caution. Be sure to check the ingredients in any pesticide you use as many have been associated with harmful reproductive effects in both men and women. The most important thing you can do is to read the label and only use the pesticide that mentions the specific pest you are trying to eliminate and only under the conditions the label describes. For example, do not use indoors a product labelled 'for outdoor use only' just because you do not have anything else handy around and are phobic about insects or rodents. (The section on agricultural workers provides more information on pesticides, and the London Hazards Centre will be able to provide detailed guidance.)

Remember that all these cleaning agents and pesticides have a greater effect on children than on adults; children under ten are the most frequent pesticide poisoning victims. Only use pesticides as a last resort, use the least toxic possible and, instead of starting a long-term spraying campaign yourself, contact a well-trained, reliable professional to do the job for you quickly and effectively.

Before you start using any pesticides you may want to try some of the suggestions below first.

Less Toxic Ways to Get Rid of Household Pests

- Insects have a keen sense of smell and avoid sharp scents like mint, tansy, basil, cedar and camphor. You can buy little jars, pierce holes in the lids, fill them with selected herbs and place them in closets, drawers and shelves.
- Make simple traps and fill them with food harmless to humans but harmful to the pests. Meat or syrup can kill flies. Beer finishes off cockroaches. Silverfish die from eating flour. Also spring traps coated with peanut butter attract different types of rodents.
- Spread technical (blue coloured) boric acid bought in hardware shops (not to be confused with medicinal boric acid) behind the refrigerator, under the sink and in corners to kill cockroaches. Vaseline barriers along the woodwork and countertop wall edges might decimate an ant invasion as a temporary solution. For long-term control, trace the invasion back to the points of entry and fill the openings with petrolatum, putty or plaster.

INDUSTRY AND MANUFACTURING: THE BLUE COLLAR APRON

Women work alongside men in virtually every industry. The number of women in each specific occupation that used to be considered male territory may be comparatively small, but together with women in traditionally female manufacturing jobs, millions of women are involved. They risk potential reproductive injury every day.

> There is plaster dust and fumes from patination of metal, ammonium sulfide, fumes from other people welding. I avoid doing welding. I miscarried during a previous pregnancy. I did welding then. The miscarriage wasn't caused by the welding as they found it was due to a blighted ovum. There are light waves given off by the welding and the boys are worried about it affecting their parts, but it didn't seem to have affected them.
>
> Gemma, US sculptor casting in a bronze foundry

> The chemicals there are real strong and I believe that by inhaling them they could give my baby brain damage or something. Or maybe the chemicals there are not good for your skin, and they are not good to be inhaling. They had masks, but they didn't give them to you. Sometimes they didn't even have the masks, so you would breathe in this stuff the whole night. But it was dangerous, I believe, to an unborn.
>
> Carla, US car seat fitter

> There was solder smoke. They said that was not really good for you by itself and worse being pregnant. We were supposed to get safety glasses and a mask, but we never got them. When you cut the wires for soldering, they can pop up and hit you in the eye.
>
> Marcy, US solderer and wirer

Altogether more than 50 chemicals commonly used by workers have been shown to impair reproductive health in animals. We do not yet know how many of these adversely affect humans too. Men as well as women face these risks and both sexes should unite to eliminate these hazards.

Four chemicals known as glycol ethers may pose reproductive and other health risks to thousands of workers. About 90 per cent of those at risk are painters, printers, furniture finishers, auto body workers and wood and metal workers. Possible damage to testicles, reduced fertility, foetal abnormalities and nerve or bone marrow damage can occur at currently allowable exposure levels. Water-based paints can be substituted for paints containing organic solvents and the printing industry is experimenting with vegetable oil-based processes.

A few industrial jobs are also thought to be associated with increased maternal risk for spontaneous abortion – those involving lead smelter emissions, the manufacture of oral contraceptives, high dose radiation, and 'industrial chemical processes'. The risk of increased numbers of reproductive difficulties also can occur through paternal exposure to certain substances. Studies have shown that partners of men exposed to vinyl chloride, DBCP (dibromochloropropane), anaesthetic gases and chloroprene have had trouble conceiving, or had increased rates of miscarriages and birth defects. Lead is the best known of the industrial reproductive hazards. Even though a blood lead level low enough to prevent harm to sperm development in males is similar to the level which will prevent damage to foetal development in pregnant women, the effect of lead exposure on men's reproductive health is often not considered.

The Meat Industry: Become a Vegetarian!

In terms of overall hazards, the meat industry is a dangerous place to be employed. Pregnant women have to work under chronically hazardous conditions. They endure extreme heat or refrigerator cold. Grease and blood make the floor and tools slippery, and the stench from open animal stomachs and bladders periodically infiltrates the air. In meatpacking plants, the roar of machinery is constant, the speed of the assembly line is fast and often increasing, and workers monotonously and repetitively hack away with knives and power saws, regularly injuring themselves and others. Even working in a retail butcher shop is hazardous.

> I worked from 7 a.m. to 2 p.m. in a retail butcher shop in an East End market. I stood seven hours a day selling meat over the counter. A butcher shop is always a dangerous place to work. There are knives and a bandsaw. If a knife fell off the block you could lose some toes. The floors are slippery. I slipped several times, but not when I was pregnant. There was a first aid kit, but nobody was specially trained in first aid. You have to be particularly careful about not being scratched by a pork bone as you can get infected. It never happened to me, but my brother got poisoned all up his arm.
>
> No extra consideration was given to me at all when I was pregnant. I was the only woman who ever stayed there. It was a

hard job. You had to do a lot for your money, but the pay wasn't too bad. Work's work, I suppose. You always have the option not to go. The boss's attitude was that everyone is replaceable so if you didn't want to come in to work you shouldn't come in. He would just dock your pay.

Penny, worker in a retail butcher shop

Electronics – Silicon Chips: The Not-So-Clean, Clean Industry

The computer chip industry has been in the news over the past few years because of the exposure of its workers to hazardous chemicals. This industry likes to call its chip fabrication areas 'clean rooms' because dust particles are filtered from the circulating air. It was erroneously assumed that these dust-filtering systems introduced to protect the chips would protect the workers as well.

The HSE is re-examining the risk to pregnant workers in the semiconductor manufacturing industry following recent reports of miscarriages associated with work activities involving exposure to glycol ethers. It will assess recent data and plans to initiate a study of women employees in the industry. The HSE will also issue information to the industry on the control of exposure to glycol ether which is covered by the COSHH regulations.

Women, who comprise a large segment of the workforce in the microchip industry, have been complaining about a wide range of work-related health problems due to their exposure to toxic gases and chemicals. These range from headaches, dizziness, breathing difficulties, allergic reactions, nausea, sore throats and skin irritations to fatigue and extreme chemical sensitivity. In the United States, it wasn't until some of the male workers developed similar symptoms and non-company doctors investigated the women workers' complaints that it became evident that the 'clean' industry was in fact very dirty. Recent studies by the American semiconductor industry have linked work in this industry with an increase in miscarriages.

When I was working at —— there were pregnant women there and they were never told of the risks of the chemicals that they were working with. If I still was there – in fact that's when I did get pregnant and didn't know it at the time – I would be very aware of what the risks are, as far as the chemicals. I would have tried to get out of that department or I would have had to leave work because it is pretty dangerous. But they never said anything.

Carol, US microcircuits assembler

The Garment Industry

Over one million British women work in manufacturing, many of these in the garment industry. This is an industry that has always depended on female labour and its history has not always been an honourable one. Although the British garment and textile industry is in bad shape and has lost 300,000 jobs since 1979, it still employs about 430,000 people (70 per cent women), is the fifth largest manufacturing industry in the UK and the second largest in Europe.

Working conditions are still poor. The GMB, the national union, averages twelve industrial injury claims a day against clothing and textile employers. Many accidents stem from poor ventilation, inadequate guards on machines and blocked access ways; sewing and cutting machines often produce high levels of noise and vibrations while poor ventilation and hot pressing machines can lead to heat stress. In addition to these discomforts, pregnant workers who sew and stitch may be exposed to hazardous chemicals in dyes, synthetic fibres, fabric treatment processes and, especially, cleaning solvents.

If you sew by hand or by machine, you may be paying with your health for consumers' thirst for easy-to-care for products. We all own garments that bear the labels 'colour fast', 'permanent press' and 'water repellent', but few of us are aware that hazardous chemicals such as biphenyl, acrylic latex, chromium and methyl-butylketone are involved in the types of processes that make these possible. Recent research has also linked three categories of dyes – benzidine, o-toluidine and o-dianisidine – to liver damage and bladder cancer.

The most widely used cleaning solvents used to remove dirt, oil and grease from fabrics (perc – perchloroethyolene; TCE – trichloroethylene) have been found to cause cancer in laboratory animals and, if absorbed through the skin or inhaled, can cause serious eye, liver, kidney, heart and nervous system damage. Many of these chemicals may also be capable of causing reproductive harm. Inhalation of synthetic fibre or cotton dust usually leads to respiratory difficulties affecting your health during pregnancy. If severe enough, this can hamper your ability to deliver a sufficient supply of oxygen to the foetus.

Women in the garment industry suffer from repetitive strain injury (RSI) resulting from repetitive movements of hands, wrists or fingers in the same position. Sitting over a sewing machine or standing over a presser for long periods of time can cause muscle strain; lifting heavy boxes is hard on your back and shoulders. The speed-ups, repetition, heat and noise combine to produce the stressful working conditions typical of the garment shop.

Stay informed. Ask your safety representative to check on whether you are exposed to any of the toxic chemicals. Use the MHSW and COSHH regulations to protect yourself.

In addition to the need for a healthier work environment, decent sick pay and pension schemes and equal pay and opportunities for women need to be implemented. Very few workers are covered by sick pay schemes and the statutory sick pay is very low. Many women working on the factory floor make little progress beyond these jobs. The disparity in pay between men and women in the industry is due to a combination of factors – differences in overtime and bonus payments and gender bias in types of jobs given to women. As with the other sectors that employ many women, improved maternity benefits, childcare and training for returners are sorely needed.

THE OFFICE: HIDDEN DANGERS

If you are an office worker, school teacher or homemaker, you may suffer from more workplace hazards than you think. Stress and indoor pollution are the two main culprits, both hard to pinpoint and hard to ignore. Indoor air pollution has reached crisis proportions and millions of women of childbearing age spend large portions of their lives in these contaminated environments. Even sitting in one place for a long time can cause harm.

If standing all day at work in an overheated factory causes tiredness of the muscles and also varicose veins, prolonged sitting may be just as harmful, for the lumbar region of the spinal column becomes bent, the movements of the abdominal viscera are interfered with, the lower ribs are compressed, and since deep inspiration is hardly possible the lungs are badly ventilated and the aeration of the blood is imperfect.

Until recently women in white collar positions who complained of headaches, dizziness, coughs, colds, nausea, generalised weakness, joint and muscle pain and various allergic reactions were thought to be 'hysterical' – a term often applied to women when those in charge cannot isolate an 'objective' cause. Now it is known that stress and ecological illness (chemically/environmentally induced disease) may account for these symptoms. Office work is a 'high-stress' occupation and pollutants in air-tight office buildings appear to be a major cause of ecological illness.

Formaldehyde, for example, is considered to be a prime culprit of allergic sensitivity as well as a carcinogen and possibly a reproductive hazard. It poses a great potential threat to our health as it is omnipresent – found in the pressed wood products which make up a large percentage of furniture and building materials. Even harder to avoid is its presence in some office supplies, tobacco smoke, perfume, toilet paper, permanent press clothing, carpeting and curtains.

Unless a building is properly ventilated, pollutants build up to levels that are unhealthy and possibly dangerous to you and your unborn children. Because you hyperventilate during pregnancy in order to obtain the oxygen you need, you also inhale more pollutants along with the oxygen. You may only be aware of minor symptoms and not aware of the possibility of more serious foetal effects.

Sometimes the ventilation systems in closed buildings primarily circulate 80–85 per cent of the same air already polluted with tobacco smoke, micro-organisms, chemicals from copying machines, cleaning fluids and synthetic clothing, furniture and building products. This is mixed with 15–20 per cent of fresh air – which may also contain pollutants. Furthermore, unless the system is correctly designed and in very good working order, it will not be able to eliminate even moderate amounts of pollutants. Because you cannot open the windows, you are captives of the heating and air conditioning engineering control systems which do not manage to keep workers comfortable much of the time. Offices, factories, hospitals, schools and homes can be plagued by tight-building health hazards.

> From what I understand we work in a condemned building. The engineer said that there was a crack down the main seam of the building. But there have been no cave-ins. But the fire department, I believe, won't pass it for inspection. But we're still working in it. There was bad air circulation. When you are pregnant the baby cuts your air anyway especially the bigger you get. Then if the building you're working in is not really properly ventilated, it puts a strain on you more than you realise.
>
> Daisy, US office clerk

> The air was very bad. In the winter and the summer or all year round it was very bad. It seemed like you were suffocating. Some days would be really hot and it seemed like I was going to pass out. This was in the summer time. We don't have any windows at all.
>
> Judith, US clerk-typist

> We have a no-smoking policy, but two colleagues smoke and management doesn't do anything even though we complained. They would rather that we fight it out. We can't open the window. When we put the fan on, I get cold. The man in the next office smokes and he puts on an ioniser that doesn't do any good and he sprays the office and it smells like a disinfectant. The man that smokes is loud and noisy. He is PR and is always on the telephone. The more you become pregnant, the more sensitive you become. It's only two weeks before I leave so I'm grinning and bearing it.
>
> Amira, librarian

VDUs: Stress on the Work Front

In addition to indoor air pollution, the automation of the office has had considerable physical and psychological effects on office workers and has received an enormous amount of media attention. Millions of people, from clerks to chartered accountants and stockbrokers, many of them women of childbearing age, make their living doing work in offices. It is estimated that approximately 10 million VDUs are being used in UK workplaces. For those of you whose jobs are varied and only use VDUs part time, many of you will probably have found them a blessing, simplifying the tedious aspects of your jobs and increasing the time spent on the more creative or decision-making aspects of their work. But most women do not hold these kinds of jobs. They, instead, have had pleasant working conditions changed to white collar assembly line positions.

Such work can now be alienating and demeaning. The VDU is master. It can automatically pace your work by bringing new tasks on to the screen and it can supervise and monitor tasks by registering mistakes. The combination of high demands with low control is the hallmark of very stressful jobs. Nowhere is this more true than with clerical VDU workers. In fact a NIOSH survey of clerical video display operators in the United States found high levels of anxiety, depression, confusion, fatigue and health problems. They suffered from even greater work stress than air traffic controllers, who at least deal with potential life and death situations which are of public concern.

> I work from 8.15 to 6.15 and I get a quarter of an hour tea break in the morning and an hour for lunch and a quarter of an hour break in the afternoon. I am supposed to get the break at 10 a.m., but sometimes it doesn't happen that way and I have to work 3½ hours in a row. That's hard. I get a pain in the back of my neck and it's hard on the eyes because you have to concentrate so much.
>
> There's no union. I know, because I asked about it. I don't know who to ask about not getting the breaks on time. It's not so bad for me as I only work part time – Monday, Tuesday and half day Wednesday – but some of the girls work 39 hours a week.
>
> Fiona, supermarket checkout scanner

VDU Complaints

- Miscarriages, birth defects, problem pregnancies
- Soreness, itching and general discomfort of the eyes
- Pains in neck, back, shoulders, arms and fingers
- Headaches
- Blurred vision and cataracts
- Dizziness and nausea
- Rashes
- Stress-related illness and discomfort
- Irregular menstrual periods
- High blood pressure
- Inability to sleep
- Tension and fatigue
- Ulcers
- Inability to relax without drugs, cigarettes or alcohol
- Poor appetite or incessant eating
- Depression
- Frequent sickness
- Chronic anger

Glare reducing devices, moveable keyboards, green coloured screens, special glasses and more frequent breaks are now being tried as remedies, and specially designed exercises performed at the workstation reduce stress and strain – ask your GP or safety representative about these. Not all physicians do know about them, but sometimes computer and software manufacturers themselves publicise the information. The exercises promote relaxation and flexibility and can easily be performed in the early months of pregnancy. It would, however, require the skill of a contortionist to do the 'knee kiss' or 'windmill' in your ninth month so use your judgement and eliminate those specific exercises that might cause you to strain yourself. If you are starting these exercises for the first time when you are pregnant, check with your doctor before you begin just to be sure that you do not have a condition that contraindicates their use.

VDUs and Reproductive Hazards: Fact or Fiction?

Most of us have read articles about clusters of VDU operators who had miscarriages, premature births or infants with birth defects. So far scientists have not been able to determine any direct causal connection between VDU use and reproductive harm; some view these clusters as clues to a serious problem while others consider them to be due to chance. The latter group reason that miscarriages are such common events that such clusters could occur by chance alone in the large numbers of women of childbearing age working on VDUs all over the world.

One study conducted at the Kaiser-Permanente Medical Center in Oakland, California found that women who used VDUs for more than 20 hours each week in the first three months of pregnancy had nearly twice as many miscarriages as women working at the terminal for under 20 hours. The researchers also found that heavy VDU users were more likely to give birth to babies with birth defects, but the increase was not statistically significant. However, the members of the Kaiser-Permanente research team emphasise that they cannot be sure that the rise in miscarriages and birth defects was due to the computer itself, or to some physical or stressful aspect of the workplace, or even to some socioeconomic differences between those who use the computer most of the time and those who use it less frequently.

In 1994, after reviewing nine studies involving pregnant VDU workers, the British National Radiological Protection Board issued a report which concluded that pregnant workers working with VDUs were not likely to have an increased risk of miscarriage or increased risk for bearing a child with congenital abnormalities.

Some workers have not been satisfied with the negative or ambiguous findings of previous research and, primarily as a response to worker pressure, scientists are still investigating low-level radiation effects of VDUs. Some experts suggest that low-level electromagnetic radiation may alter or disrupt embryonic or foetal cell development. Experiments with mice and chickens have shown these effects. Other scientists claim that the designs of earlier studies underestimated the link between VDU use and pregnancy complications. Liquid crystal screens are installed in laptop and notebook computers and researchers are working on the design of video display screens made from different kinds of material. Researchers in several countries are carrying out studies that compare a group of VDU workers with a control group of non-VDU workers with similar characteristics to determine whether there is a definite relationship between VDU work and reproductive difficulties. A large number of cases will have to be studied over a long period of time before any firm conclusions can be drawn, but in the meantime those workers who use VDUs intensely for the entire work day are considered to be at greatest potential risk.

The four areas of most concern are:

- exposure to electromagnetic fields (EMFs)
- poor workstation ergonomics
- stress associated with work overload
- lack of variety and control over work.

It may be a combination of all of these which explains the lack of evidence implicating any one condition as a reproductive hazard.

Information about foetal damage from other sources of radiation indicates that the most critical period for possible harm occurs during the first five weeks of pregnancy. Furthermore, the potential risks of VDU work to men have not yet been identified. Given this amount of uncertainty a policy that allows both men and women trying to conceive to transfer temporarily to other clerical jobs without loss of pay and status should be initiated. Some unions have followed this line of reasoning and have negotiated agreements with employers giving pregnant women the right to perform other equivalent work not involving the use of a VDU during pregnancy, or to take pregnancy leave without loss of seniority rights.

I don't spend too much time on the video display unit. If we get too much, we can pass it on to the word processing department. My boss has treated me very well. He has five children. I think I've just been lucky.

Diana, secretary

The London Hazards Centre has published a book on VDU work: *VDU Work and the Hazards to Health* (London Hazards Centre 1993). Until answers to questions about reproductive hazards and VDU have been decided, it suggests the following:

London Hazards Centre Suggestions for Protecting Against VDU Pregnancy Hazards

- The best protection is to gain the right to transfer to non-VDU work when trying to become pregnant or when pregnant.
- Take regular breaks from VDU work – preferably at least ten minutes per hour.
- If possible, limit your VDU work time to half a day or less – problems seem to be associated with increased time spent at the machine.
- Negotiate for VDUs that comply with Swedish guidelines for emissions or for liquid crystal display units.
- Switch off the VDU when it is not in use.
- Sit as far way from the screen as is compatible with visual and physical comfort. You receive more electromagnetic radiation emissions when you sit close to the screen.

If you have or know someone who has a VDU related pregnancy problem, contact City Centre, the London Hazards Centre, the TUC or the HSE (all addresses are listed in the Useful Addresses section at the end of the book).

In order for you to make up your own mind as to how much of a risk your individual VDU work situation poses, it is useful to understand the components of a VDU, how they operate, what the known adverse health effects are and the basis for suspicions of possible additional types of harm. Remember that being in good physical shape during pregnancy, as well as avoiding specific reproductive hazards, is important for your own health and that of your unborn child.

Heat

VDUs produce about 300–400 watts of heat output – somewhat more than electric typewriters. This heat is not a hazard by itself, but large numbers of VDUs combined with poor ventilation can raise room temperature to a level of discomfort. Because you have to cool your foetus as well as your own body, you will be made even more uncomfortable when you are pregnant. Heat output should be measured near the screen as well as other places in the environment as mucous membranes in your nose and eyes may dry out if the surface temperature of the VDU is too high. Fans can be used to ventilate VDUs, although they can cause unpleasant draughts which result in muscular pains. Check the temperature on the surface of the VDU after two hours of continuous operation. If it exceeds 90 degrees fahrenheit or 34 degrees celsius, bring it to the attention of your supervisor.

Noise

Noise from VDUs comes from ultrasound or high-pitched noises from electrical components, the fan, disk drives, printers and other adjunct equipment. This noise is seldom hazardous to hearing but it can be annoying and therefore stressful.

Radiation

The cathode ray tube can produce a certain amount of ionising radiation (X-ray) and various electronic components can emit non-ionising radiation (radio frequency). The ionising radiation, potentially the most dangerous hazard, is almost entirely absorbed by the thick glass screen. The non-ionising radiation can be greatly reduced by the use of a simple metal shield. You should press for the installation of these shields on the current generation of VDUs. Some health scientists are urging the development and use of a non-cathode ray tube VDU. Non-ionising radiation has been inadequately studied in humans, but harmful reproductive effects have been found in animal studies. A 1991 Finnish review of experimental studies found that the type of magnetic fields which are associated with VDUs may have an effect on the embryos of mice, rats and chickens and that the effects of the magnetic fields on these animals were greatest during the very early stages of development.

The Effects of Non-Ionising Radiation

High dose microwave and radio-frequency radiation (in animal studies)

- Chromosome damage
- Reduction in sperm count
- Cataracts
- Blood and nervous system disorders

Low dose microwave and radio-frequency radiation
- Lowered fertility, death of embryos, increase in foetal malformations and stillbirths in animals

High dose ELF (extremely low frequency radiation)
- Lower birth rates
- Birth defects
- Female reproductive disturbances
- Increased offspring death in animal studies

Low dose ELF
- Abnormal embryonic development

So far tests in the US, Canada and the UK show that the radiation from VDUs is extremely low – usually far below the acceptable occupational health level. While the results of these tests suggest that there are no significant radiation hazards from VDUs, there is some reason still to be cautious. Older VDUs, VDUs that are badly maintained and VDUs not made according to strict safety standards in force in most industrialised countries may not be as safe as those tested. The occupational exposure levels also may be too high and the effects of long-term exposure to low levels of non-ionising radiation are still mainly unknown.

More important, as the long-term effects from exposure to low levels of non-ionising radiation are unknown, there are currently no standards to protect workers from non-ionising radiation in VDUs – ranging between radio frequency, very low frequency (VLF) and extremely low frequency (ELF) radiation. Most of the non-ionising radiation emitted by VDUs occurs in the VLF range. Almost no studies have been done at these frequencies, but biological effects have been found at both higher and lower frequencies.

A report from the National Radiological Protection Board (NRPB) stated that no clear evidence exists of a cancer hazard from electromagnetic radiation produced by VDUs or overhead power lines, but also stated that the existing evidence does not exclude the possibility either. The report urged more research to clarify the possible links between cancer and low frequency electromagnetic fields in the home and at work with particular emphasis on comparative exposures of men and women in different occupations.

Summary of VDU Health Studies Recommendations

- VDU equipment should be manufactured or refitted with protective metal shields to reduce radiation emission.
- VDUs should have adjustable screens and keyboards with glare reduction devices.
- There should be mandatory, periodic testing of equipment.
- Workers should have scheduled rest breaks – fifteen minutes off for every hour on the machine – and do a maximum four to five hours of VDT work a day.
- There should be an elimination of machine-pacing or computer-monitoring of workers' output.
- There should be mandatory visual testing of operators.
- Workers should keep track of symptoms they and co-workers have and record all incidents of machine breakdowns.
- There should be worker participation regarding the use of VDUs in issues including job security, workstation design, maintenance, job rotation and any performance monitoring programmes.
- Ergonomic requirements should be as follows:
  - an adjustable table that allows leg movement and at least seven inches of knee clearance
  - a foot rest for short operators
  - a simple, movable document holder
  - a maximum eyes to screen viewing distance of 27.5 inches.

The Health and Safety (Display Screen Equipment) Regulations came into effect on 1 January 1993 and cover most of the above recommendations. (Further information about these regulations can be found in Appendix C at the end of the book. Also see the section on musculoskeletal injuries, earlier in this chapter, for material about repetitive strain injuries.)

Other Office Equipment

Next to VDUs, photocopying machines are the main staple of modern offices. When the photocopying machine breaks down, so does the office system. Some people spend much of the day photocopying memos and reports; others use it only occasionally. What can be harmful about such a work-saving machine? Surely it is better than using reams and reams of carbon paper? True, but there are hazards with copiers too, particularly if they are old, not maintained well and are run continuously in an inadequately ventilated space. Photocopiers usually operate by an electrostatic process and emit ozone, a sweet-smelling gas formed from the oxygen in the air when it has been energised by the high-voltage source in the machine. Ozone can impair your lung function which in turn can reduce your resistance to disease. It is also suspected of being a mutagen causing genetic damage. In light of these suspicions it makes sense not to sit in the path of the photocopier exhaust.

The principal solution to the ozone problem as well as other indoor air pollution problems is better ventilation. Most manufacturers have minimum space and ventilation requirements for the machines they sell. Check the manual for the equipment you have or call up the rental company for this information. If your location falls below the standard required, either move your machine to a safer location or rent a different type of photocopier that fits your environment.

Another hazard of photocopying machines is the chemical composition of the toners used in the machines. Exposure to users is usually minimum but could be serious for maintenance personnel. If you are responsible for changing the toner, do not touch it with your bare hands or breathe it directly; some chemicals previously used in toners have been found to be mutagenic in bacterial studies and there is a risk that these substances could cause cancer or reproductive harm in humans as the DNA in bacteria is the same DNA in humans. Be extremely wary of any substance that is found to be mutagenic in animal or bacterial studies until they are proved harmless to humans.

Carbonless carbon paper, another boon to the office worker, also has its dark side as some have formaldehyde as one of its ingredients. A few workers become sensitised to formaldehyde and others find the paper irritating to their nasal membranes, bronchial tubes, and skin.

THE SERVICE SECTOR: HIGH STRESS, LOW WAGES

Many new jobs for both men and women are now in the service sector. Some of these are in higher-paid management positions and women are beginning to climb the corporate ladder. If you are one of these women you are likely to be able to exert greater control over your life both inside and outside the workplace and, although you may suffer from many of the same problems as women at the lower end of the pay scale, it is often to a lesser extent. You too though face a great deal of stress, are expected to work overtime and often are hesitant to ask for special consideration if you are pregnant or have family responsibilities.

A great many women however still occupy the lower positions in the service sector and may suffer more from working conditions that may be harmful to their health, particularly when they are pregnant. Many of these lower-paid occupations within the service sector have many of the same characteristics. The jobs are often high-stress and low-paying, have little prestige and no control, are non-unionised, may be part time or involve evening or weekend hours, may require workers to stand for most of the day, expose them to customers' infections, require them to appear cheerful and friendly at all times – and are filled by women. Many of these women are of an age where they are contemplating pregnancy, are pregnant or have young children.

In addition to all of the above, some of the service sector jobs subject workers to chemical hazards, indoor air pollution, extremes of temperature, noise and vibration, heavy lifting and carrying burdens and physical and social stress.

> There is a lot of bending and working in reasonably odd positions. I work with some machines, but it is mainly manual. Carrying buckets and working with large moulds puts a strain on the body. I visited an osteopath before I became pregnant. The workshop is very noisy, very cold in the winter and very hot in the summer. In the winter, it gets very hot when we are heating metal. The workshop is as safe as it can be, but it is inherently dangerous when it deals with fire and molten wax. I get sufficient breaks and if I wanted to sit around for longer, I could.
>
> Gemma, sculptor casting in a bronze foundry

Violence in the Service Sector

Large numbers of women in the service sector work in establishments that are prone to violence, such as eating and drinking places, hospitals and shops. A few examples of workers who may be subjected to violence at work are: nurses, teachers, utility workers, social workers, traffic wardens, administrative workers and guards. The risk of being injured is increased if you exchange money with the public, work alone or in small numbers, work late at night or early morning, work in high crime areas or work in community settings. A combination of relatively simple changes such as separating workers from the public, installing bullet-proof barriers and alarm systems, adding additional staffing and a ban on working alone can help a great deal. A law passed in Gainesville, Florida, for example, required that two people work on an 8 p.m. to 4 a.m. shift. Not only did this result in an 80 per cent decrease in assaults, but also in increased sales as customers as well as workers felt safer. Furthermore, there are many different ways to define violence at work. Currently, a broad range of behaviours such as sexual and racial harassment and verbal abuse which can also result in physical or psychological injury are usually included within the definition of violence in the workplace. When you are pregnant, it is even more important that you seek changes if you feel that you might be vulnerable to violence at work.

Beauticians: What Price Beauty?

Beauticians and cosmetologists face all the problems of other pregnant workers in the service industries cited above, and in addition are exposed to many potentially toxic substances in organic hair dyes, hair sprays, hair straightening and permanent waving solutions, nail varnishes, plasticisers and resins and depilatories. Many beauty workers realise that chemicals that they work with may be harmful, but far fewer are aware how extremely toxic some of these products may be. In the 1970s a NIOSH study of cosmeticians in the United States exposed to hairsprays found that they were more likely to develop chronic lung disease than were a comparable group of non-exposed university students.

Organic hair dyes have also been flagged as potential hazards to those manufacturing them, those applying them and those having their hair dyed. Dr Bruce Ames, an American biochemist, tested 169 permanent hair dyes and found that 88 per cent caused mutations in bacterial laboratory cultures. Eighteen different chemicals were used in these dyes and nine were found to be mutagenic. Dr Ames believed that these mutagenic substances in the dyes were also likely to be cancer causing and could be absorbed through the scalp and skin. He recommended that a study of both birth defects and cancer be conducted among the millions of women using hair dyes as well as of workers who manufacture hair dyes or use them regularly in their jobs.

Continued standing worsens varicose veins in the pregnant worker and having your job depend in part on your pleasant manner adds an additional strain to discomforts of pregnancy, as does working very often in a poorly ventilated and humid salon. Being exposed to chemicals that are known to cause allergic reactions, skin problems and asthma may also have additional yet unknown effects.

Because toxic substances can be absorbed through the skin, be sure to wear gloves when using hair dyes, perming lotions and other products even if it is easier and quicker to work with bare hands. The long-term safety of most of these products is not known. In addition these substances, particularly those applied by spray, linger in the air and can be inhaled. Realistically, although wearing a protective mask might aid your health, it is likely to scare your customers away and cost you your job, and so less obtrusive protection methods must be found. Substitute sprays with mousses and gels which can be directly applied to the hair and which result in less inhalation of the chemicals. If you wear gloves while administering the products the skin route of absorption is eliminated.

While some of the most toxic hair dyes have been removed, the potential hazards of those remaining are unclear, as are the possible human reproductive effects of many of the solvents used in nail beautification products. A new technique of nail sculpting involving exposure to toluene, isopropyl alcohol, butyl acetate, ethyl methacrylate, methacrylic acid and other dusts may have adverse long-term health effects on both the regular customer and sculptor, particularly if she is pregnant. Methacrylate dusts are known mutagens in animals. Some of the animals injected with methacrylates have developed adverse reproductive effects, including foetal death and birth defects. Nail sculptors should work with high-powered suction ducts directly over the workbench in order to reduce the fumes and dust levels.

Dry Cleaning and Laundry Workers: Clean Clothes, Not-So-Clean Workplace

Later sections discuss the effects of excessive noise and heat and heavy lifting, all of which can lead to reproductive difficulties. Dry cleaning and laundry establishments possess the potential for all three problems. Laundry and dry cleaning machines are normally operated at high noise levels. If a worker is subjected to this level of noise for a long period of time then permanent hearing loss may be suffered. In animal studies, noise has been found to be toxic.

In addition, some chemicals used in the dry cleaning process may be hazardous. The HSE is carrying out a three-year investigation to the possible link between workplace exposure to the solvent perchloroethylene (a commonly used dry-cleaning fluid) and increased risk of miscarriages among women working with the substance.

The hot, humid atmosphere of laundries and dry cleaning plants can stress the heart and circulatory system. Dehydration, an effect that is exaggerated in a pregnant woman, is more likely to occur if she is engaged in strenuous physical activities in such an environment. Dehydration interferes with blood circulation to the foetus and may trigger premature labour. A pregnant worker should take particular care not to get overheated, as she has a harder time keeping her body temperature constant because she has to eliminate the extra heat emitted by the foetus. Overheated conditions add to body stress. The pregnant worker, after vacations and weekends off, should be sure to get used to the hot conditions slowly by reacclimatising – gradually increasing the number of hours she spends in the hot environment . (See the section on Heat and Humidity for further information.)

Lifting and carrying bundles of heavy laundry and working while standing puts a strain on the worker's back and legs. This too is to be avoided during pregnancy, as your back is already strained by the pressure on the lumbar region. Ask someone to share the lifting and try to have breaks in which you can sit for a while. Also avoid hazards such as slippery floors and escaping steam.

Preventing Hazards in the Laundry

For the employer

- Install dehumidifiers and air conditioners.
- Install sound-absorbing material on floors, walls and ceilings.
- Keep up a good maintenance schedule to reduce the noise of the machinery.
- Keep floors dry at all times – clean up spills promptly.
- Make sure all particles are vacuumed at least once each day, and more often if the workplace is unusually dusty or contaminated material is being cleaned.
- Ensure you are notified if you are sent contaminated laundry to clean and know what it is contaminated with.

- Provide a quiet, air conditioned rest room.
- Ensure that there is good ventilation and circulation of air – this is a must.

For the pregnant worker

- Spend short periods in the rest room and sit with your feet elevated.
- Have a gradual reacclimatising schedule after holidays.
- Ensure cool drinking water is always available.
- Use hand creams before, during and after work.
- Ensure you are notified if you are sent contaminated laundry to clean and know what it is contaminated with.
- Take no chances of slipping and wear shoes with no-slip soles.
- Try to alternate lifting with non-lifting jobs. Make sure that you stretch your muscles after lifting to reduce the strain on them.

Flight Crew Members

Flight crew members, particularly flight attendants, suffer from stress, changing shifts of work and time zones and exposure to infections from passengers. There is also an additional risk of being exposed to solar (from the sun) and galactic (from the stars) radiation, particularly on certain routes and during years when there is greater sunspot activity. Pregnant women on the ground are largely shielded from these two sources of radioactivity by the atmosphere, but no shielding has yet been designed that will prevent high-energy radiation from filtering into high-flying aeroplanes. In 1989, the United States Federal Aviation Administration issued a warning that pregnant women working on long-haul, high-altitude flights over polar routes would expose their foetuses to larger amounts of radiation than federal standards recommended. By mid-1990, the agency was still evaluating the evidence as to whether further protective action was needed. It was planning to issue a computer program that could be used by individuals to help figure out for themselves their radiation doses.

Pregnant flight crew members in the UK are covered by the new legislation implementing the EU Pregnant Workers Directive, but the Europilote Association wants a better pregnancy/maternity leave policy.

Europilote Association Pregnancy Recommendations

- A healthy pregnant flight crew member should be given the choice to continue or cease flying immediately.
- Pregnant flight crew members should cease flight duty at a maximum of 26 weeks.
- Information concerning possible hazards associated with flight duty while pregnant should be available to those who wish to become pregnant, to potential fathers as well as the pregnant worker.
- A returning crew member after pregnancy should be reinstated in her previous job without any affect to her seniority.
- For the periods when a pregnant crew member chooses not to fly, she should be offered a suitable ground job. If this is not available, she should be paid her full salary.

The Retail Trade: Customers Know Best!

Image is of high priority to retailers. Sloth and slovenliness tend to be unforgivable employee traits; sales personnel are required to stand when they are in public view so as to appear busy; they are required to wear designated, neat uniforms or stylish or smart clothing.

> I had to stand most of the time. Sometimes it was a problem. I got tired very easily and got lots of aches and pains. There was a chair, but we weren't supposed to sit down. There was more pressure on me because I knew more about stock than anyone else. Sometimes, when I had trouble standing, they would send me to the medical office which didn't do much good. They would let me sit down for a while and then send me back to the floor.
>
> I had to look nice and wear a dress made of thin material so I had to wear a cardigan all the time. They should have changed my job to give me more of the written work.
>
> Louise, department store sales assistant

Prolonged standing, particularly in high heels, can cause leg pains and varicose veins. Pooling of blood in the legs is a common problem for pregnant women and prolonged standing aggravates the condition. Stress, also part of the saleswoman's lot, is caused by demanding customers, low wages, lack of respect, having to work some evenings and Saturdays and the necessity of being continually on view with a smile and polite word for customers.

Continual contact with the public increases exposure to infectious and communicable diseases. Try to keep your distance from a customer – or customers' children – during the flu season or if you know there is an outbreak of German measles, measles or chicken pox in your area. You can never tell whether a disease someone is carrying may pose a threat to your foetus, although the chance of this happening *is* small.

These are the 'front of house' problems, but behind the scenes other types of unhealthy conditions prevail. In the stock room, for example, safety hazards come from boxes stored in aisles and haphazardly placed on shelves so that it is easy to trip over merchandise or have it fall down when you are reaching for something. Lifting, transporting merchandise and reaching up and down shelves causes back and shoulder strain. Often shops have no windows and you will suffer from indoor air pollution caused by tight-building syndrome, as do office workers.

Workers at checkout counters in supermarkets are increasingly at risk from injury due to the new electronic price scanners. Employees are exposed to the hazards of cumulative trauma to the back, hands and arms caused by constant and excessive twisting, stretching and lifting involved in processing items through the scanners at the checkout counter. All of these twisting and pulling motions are even harder on the body of the pregnant worker.

Preventing Hazards in the Retail Trade

For the employer

- Provide stools to allow the pregnant worker to rest periodically.
- Carpet floors – this reduces leg pains.
- Provide adequate reaching devices such as poles, stools and ladders.
- Allow pregnant workers to alternate between sitting and standing where possible.
- Provide workers with training in proper lifting techniques.

At the checkout

- Limit the extent of the arm reaches.
- Control the height of any work surface.

- Provide a 'sit–stand' stool for alternate sitting or standing to allow rest for overworked muscles.
- Install right- and left-handed checkout counters.
- Provide task rotation and work breaks.

For the pregnant worker

- Take frequent breaks – either sit or walk around.
- Wear comfortable, low-heeled, non-skid shoes with support hose.

Waitresses: Warm Food and Cold Patrons

Waitressing is one of the most stressful types of jobs and pregnant waitresses are likely to be fairly regularly subjected to both physical and psychological stress. In fact a Canadian study found that waitresses, along with nursing assistants, hospital attendants and certain types of sales personnel, have more spontaneous abortions than would normally be expected, and the researchers hypothesised that the high stress level implicit in many of these jobs is partially responsible. It is in the early months when waitresses are adjusting to the changes in pregnancy that they are likely to suffer from the pressure and work conditions the most. They are not likely to work in the last trimester of pregnancy as their bodies become too unwieldy for the tasks required.

Waitresses are frequently non-unionised and are caught between the demands of their employers and customers. They can work shift schedules or long hours – slow periods characterised by boredom combined with frenetic activity during meal times. During the latter, they run back and forth between kitchen and tables and may have to suffer the tongue lashing of ill-tempered customers and members of the kitchen staff who are themselves stressed. During busy periods there is often no time to eat. (In some establishments waitresses may be given left over food to take home – be careful with this, as it may have turned bad from standing out too long.)

Exposure to infections, extreme changes in temperature, poor ventilation and heavy lifting pose additional difficulties for the pregnant waitress. During the summer, poorly ventilated kitchens can become extremely hot while the dining area of the restaurant may be air conditioned. Waitresses then spend most of their shifts going back and forth between areas in which temperatures vary drastically (although this is not as severe a problem in the UK as it is in warmer climates).

Furthermore, many waitresses are subjected to passive smoke inhalation from customers who smoke while dining, and they can pick up infectious diseases from those who sneeze or cough into their napkins which the waitress has to pick up along with the dirty dishes.

> I worked behind the bar serving drinks. I worked from 11 a.m. to 3 p.m. I had to stand, but that really wasn't a problem as I could sit occasionally if I felt tired. There were a lot of stairs to go up and down. The bar was upstairs and the restaurant was downstairs and I would have to bring drinks to customers.
>
> There were always other people serving so I could go to the loo if I needed to. There were no breaks during the working period. I worked because I wanted to and friends owned the bar.
>
> Jenny, barmaid

Constant standing and running around and the lifting of heavy trays combine to cause back, shoulder and leg problems. Stretching exercises will help this as will support hose and well-fitting, non-skid-sole walking shoes.

TEACHING: HAZARDS IN THE SCHOOL

Teaching, particularly at the pre-school and primary school level, has traditionally been a woman's job and, to a great extent, still is. Whereas in the past, women were forced to leave their jobs either when they married or when they became pregnant, these barriers have now been withdrawn and thousands of pregnant women teach through most or all of their pregnancies at every level of the educational system. Yet, on the whole, school administrators have done little to ensure their health and that of their unborn children.

While schools are safer places to work than many manufacturing plants, teaching is not a hazard-free occupation. The London Hazards Centre reports that employees in the educational field are five times as likely to be injured in accidents at work than employees in banks and the financial industry.

Pregnant as well as non-pregnant teachers face risks from asbestos crumbling from the ceiling, discipline problems which may involve physical violence, high noise levels, childhood disease epidemics, insufficient opportunity to use toilets, old copying machines used in airless rooms, unsafe laboratory procedures, and toxic art supplies.

> There were bugs in the bathroom, a set of slippery staircases and a room that they were removing asbestos from. I had to stay out of that room. I was concerned about keeping warm and dry on recess duty. You could use the bathroom whenever necessary as long as there was a teacher available to watch the kids. Sometimes there wasn't though.
>
> Dinah, US primary school teacher

In 1993 over 11,000 accidents to students and staff were reported, and accidents have been increasing in almost every educational setting. Because the number of accidents in schools was so high, schools and local education authorities requested guidance on safety management from the HSC, which published a booklet called *Managing Health and Safety in Schools* (£5.95, available from HSE Books, PO Box 1999, Sudbury, Suffolk CO10 6FS. Tel: 01787 881165).

The Art Room

If you are a pregnant art teacher, you are likely to suffer higher risks from toxic exposures than your non-pregnant colleagues or students. Because your respiratory rate has increased you may inhale greater quantities of toxic substances from the air, so be sure that the ventilation is adequate. Substitute less toxic products whenever you can and be extra careful handling the materials. These measures will reduce the exposure level substantially. For example, common solvents such as benzene and tetrachloride used by sculptors, printmakers and painters in mixing paints, varnishing and cleaning are very toxic. Acetone or grain alcohol are good substitutes, but care should be taken as they are highly flammable.

Some paints, solders, and dyes are toxic to the kidneys, liver and lungs. Some can cause cancer and damage male and female reproductive systems. Beware of those whose ingredients include heavy metals – cadmium, chromium, lead, mercury and manganese – suspected carcinogens and mutagens. In addition clays, glazes and sculpting stones can contain silicates. Mixing, grinding and carving processes create silica dusts which cause silicosis, a lung disease.

Firing processes also can cause the formation of toxic gases. Aside from the chemical ingredients, poor maintenance and lack of safety features on equipment and insufficient circulation of fresh air can lead to increased hazardous exposure.

Preventing Hazards in the Art Room

- Read all labels carefully. Unfortunately, too many art supplies provide insufficient information. Don't take any chances, particularly if you are pregnant. If in doubt contact your safety representative, union or the London Hazards Centre advice line (0171 267 3387).
- Make sure all containers are tightly covered and stored in safe places.
- Wear protective clothing and goggles when required.
- Vacuum dust and particles. Sweeping only spreads them in the air and onto your clothes.
- Do not eat, smoke or drink in your work area.
- Make sure that the art classroom is properly ventilated.
- Do not mix liquids while disposing of them. Throw away rags and towels used for cleaning. Do not leave them around or in the waste paper basket.
- Follow these rules at home as well as in the art room. The same toxic ingredients in inadequately ventilated rooms, improperly handled, will cause the same harm regardless of location.

Science Laboratories

Science laboratories are another hazardous area for the pregnant teacher. In school labs, the hazards of working with students are substituted for the hazards of working with patients. Students tend to be nonchalant about safety precautions. Potentially explosive, carcinogenic, toxic, and irritant chemicals are being used and often students, and less qualified teachers, are unaware of proper precautionary procedures. You are protected under the COSHH regulations and you should receive information and training about the safe use of hazardous chemicals. If you are not, speak to your union, safety representative or the HSE.

Along with being adequately informed about the toxicity of the agents and chemicals used in the labs, make sure you (and the students) wear protective clothing and closed shoes.

Duplicating Machines

Duplicating and copying machines in the staffroom pose the same hazards that they do in the office (see the earlier section of this chapter on office machines). Teachers sometimes use mimeographing and duplicating machines requiring stencils and inking of drums because they are much cheaper than Xeroxing. These, however, aggravate the problem even more as the machines tend to be older and may not be regularly maintained, leaving users having to poke and pull in order to get the mimeographing machine operating or the duplicating machine with the smelly, wet copies to print clearly. If you do have to do this, don't rush. You are more likely to get chemicals all over your hands if you are rushed.

Still more dangerous is that these machines are frequently located in the staffroom where you and your colleagues rest and eat. Thus you are apt to munch while you copy, ingesting the chemicals along with your lunch and adding pollutants to colleagues' lunches as well.

> There was a teachers' lounge which also had the Xerox and the ditto machine. We ate in there and any lounging was done in that room.
>
> Kathy, US second grade teacher

Infections

Any job that brings you in close working contact with the public brings you in contact with many different kinds of germs. Pregnant teachers who work with younger children are particularly at risk because this is the age group that becomes infected with all the childhood diseases. Those who evaded the contagious diseases when they were children may catch them from their pupils, and adults usually have more serious cases. For example, German measles (rubella) is a known human teratogen, and all female teachers contemplating pregnancy in the future should be vaccinated against the disease if they do not already have immunity due to previous exposure. Male teachers who plan to start a family or have additional children are also vulnerable; severe cases of mumps can cause sterility.

Pets can also breed disease and pregnant teachers should be wary of sick hamsters and gerbils. In addition, with more mothers working, more children are being sent to school sneezing and coughing whereas in former days they would have been kept at home. Most diseases are infectious just before the child becomes sick and this is the time the child is likely to be in school. Mothers become skilful at knowing when their children are faking, but they are not infallible in separating complaints of not feeling well due to illness and those due to not wanting to go to school.

Because more and more mothers are working, more children are attending daycare centres, and more daycare personnel are needed. This new cadre of women are even more at risk of stress overload and infections than nursery school or primary school teachers. These very young children may not be toilet trained nor have they yet developed good cleanliness habits. Diarrhoeal infections, including the parasite giardia lamblia, hepatitis-A, upper respiratory infections, rubella and cytomegalovirus (cmv) have been known to spread through daycare centres.

Over-exertion from lifting is another chronic hazard for the pregnant daycare or nursery school teacher. Lifting heavy children puts just as heavy strain on your back as the heavy lifting of objects.

> I lifted 35-pound children and retrieved kids that climbed too high on the playground equipment. There was a lot of bending and from September to November I had contractions every time I would bend over. The kids would just wait for them to finish and then they could keep going.
>
> I had no break for a five-hour period. There was no snack bar with nutritious food. We just had whatever we fed the kids which was not necessarily nutritious. There was a kitchen that we could use where we could store sodas or juice, but we couldn't have it until lunch or work was over.
>
> If we had a circle to read the children a story, we had to be on the floor with them. At the end of my pregnancy, I asked if I could please use a chair because to get down on the floor and to get off the floor, stand up and down, jumping around when you are eight months pregnant was really hard. They told me begrudgingly that

> I could do that. If I had time to actually sit and elevate my feet that would have really helped.
>
> Amy, US daycare teacher

General Hazards

HEAT AND HUMIDITY: IT'S TOO HOT IN HERE!

Workers in bakeries, canneries, laundries and garment and textile factories frequently work in hot and uncomfortable environments. You may feel ill or dizzy when exposed to such extreme heat and humidity at work.

> For the first two weeks of my pregnancy it was 140 degrees up here and I was getting dizzy. I couldn't figure out why because I'm so used to the heat. So I knew I had to be pregnant.
>
> Betsy, US finisher and presser in a dry cleaning establishment

If these conditions are severe or prolonged, you may suffer from heat stroke and heat exhaustion as well as experiencing increased fatigue and discomfort.

Sometimes the specific heating system for any type of workplace can cause problems.

> It was always too hot and nothing could be done. I asked the headmaster and he explained that the school was part of a block of council flats and they all were heated by the same system. Opening a window was not a solution because you got a draught. Sometimes the heat became unbearable. There was a large change in temperature when you had to go out in the cold. The head of the school was very nice and cancelled my playground duty.
>
> Hannah, part-time teacher

Pregnant women are more sensitive to high temperatures because their bodies have to rid themselves of heat produced by their own increased metabolic rate plus displacing the foetal body heat which is about one degree higher than their own.

If while pregnant you engage in vigorous activity for more than 15–20 minutes at a time your body temperature may increase. This usually doesn't cause problems, but if it occurs frequently, particularly when the weather is hot and humid or when the workplace is hot and stuffy, your unborn child could be harmed. The foetus does not have the capability to cool itself and your overheated body will not be able to accomplish its normal task.

Dehydration is an additional hazard if jobs require strenuous activity in a hot environment, as is the loss of body salt. You can prevent this by drinking large quantities of water or other fluids and limiting or spacing out your activities. Less body heat is produced during intermittent high periods of activity than during steady heavy work. Dehydration can interfere with the amount of blood the foetus receives and can trigger early labour. Increased demands on the circulatory system can impair alertness, mental functioning and physical capabilities. You will feel particularly uncomfortable when there is an extreme temperature change – try to avoid work situations involving drastic temperature shifts such as occur when entering and leaving cold storage rooms or hot furnace areas.

Your employers can protect you by gradually building up the time you spend under the high temperature conditions. They can provide properly designed clothing that permits air to circulate, adequately ventilate the work area and enclose the heat producing operations as much as possible.

On 1 January 1996 the Workplace (Health, Safety and Welfare) Regulations 1992, covering temperatures, ventilation and lighting came fully into effect. These regulations replace sections of the Factories Act 1961 and the Offices, Shops and Railway Premises Act 1963 and cover all workplaces except for ships, moving vehicles, mines, quarries and building sites. As relatively few pregnant women work at the exempted sites, these new rules should cover most pregnant workers. Regulation 7 covers temperature; *The Daily Hazard* No. 30 published by the London Hazards Centre provides information about hot working conditions (contact them at Headland House, 308 Grays Inn Road, London WC1X 8DS. Tel: 0171 837 5605).

LIGHTING AND GLARE: VISION PROBLEMS ON THE RISE

Lighting problems are not thought to affect the reproductive functions of workers or the health of the foetus directly. Indirectly, improper lighting can cause a great deal of discomfort for the pregnant worker as light affects comfort, safety, efficiency and mood as well as ability to see. Both extremes – too little light as well as glare from light – cause difficulties. Eye strain from VDU work, a universal complaint, has been documented in many studies. In fact, a majority of women in a survey taken by 9–5, the National Association of Working Women in the United States, responded that lighting was the most important physical aspect of their workplace. Not only does inadequate or improper lighting permeate the environment, but it is usually an indication of generally poor physical working conditions.

A German study on the impact of the office environment on the health of office workers found that VDU users reported more health disorders than non-users. All health complaints increased when the workstations were further from the window. In addition, the type of artificial light in the office played a major role in the complaints. Overhead fluorescent lighting was considered to be the worst kind while a mix of indirect general lighting and desk lights that could be focused on the task was thought to be the best.

> We have fluorescent light. I prefer natural light. Fluorescent light sets off migraines for me, but I haven't had one since I've been pregnant.
>
> Amira, librarian

After new lighting was installed in two of the original organisations participating in the study, respondents reported significant improvement in their health and well-being. The precise amount of light we need depends on our eye sight, age and the task we are doing; the finer and more detailed the work, the higher the level of light needed. For example, performing intricate tasks under the microscope producing microchips and circuits in the electronics industry, or performing close, precise bench work tasks, may cause you to hunch over into a position that puts pressure on your neck, particularly in the later months of pregnancy when the centre of mass of your body has shifted forward. This is made worse when either bifocals are worn or there is inadequate lighting causing eye strain which then spreads to the head and neck compounding pain and discomfort. Whenever vision correction is needed for this type of work, intermediate or near vision glasses rather than bifocals may be better.

The effect from illumination glare requires further research. Glare is caused by light shining directly into the eyes or bouncing off a surface and reflecting into the eyes and can be reduced by shielding the lights with plastic or relatively inexpensive devices called diffusers that spread out the light. Glare can be further reduced by eliminating shiny, reflective surfaces on furniture and equipment. Reducing glare from the VDU screen has led to the manufacture of auxiliary devices and glare-free screens have become a selling point in the highly competitive VDU market (the section on VDUs provides further details about the hazards of VDU work and what to do about them).

The Eyecare Information Service, a non-profit making body, publishes a free leaflet called *Your Eyes and VDUs* which includes information about lighting, an aspect of the VDU workplace that sometimes receives inadequate attention. Inadequate lighting and uncomfortable temperature levels can exacerbate eyestrain, headaches and mental distress as well as musculoskeletal difficulties. (Send a stamped, addressed envelope to VDU Leaflet, Eyecare Information Service, PO Box 3597, London SE1 6DY.)

In order to conserve energy or to save money, employers frequently reduce lighting levels. Poor maintenance is also responsible for dim lights; light bulbs that are not dusted can reduce the light level by approximately 5 per cent a year. If this is a problem in your workplace, suggest to management that they could clean fixtures, replace defective equipment and filter air to reduce indoor pollution. These relatively simple and inexpensive measures may correct the condition completely, and at least will increase the brightness of your surroundings.

The Workplace (Health, Safety and Welfare) Regulations 1992 state that lighting must be 'suitable and sufficient'. The approved code of practice which accompanies the regulations states that 'lighting should be sufficient to enable people to work, use facilities and move from place to place safely without eye-strain'.

MUSCULOSKELETAL INJURIES: LIGHTEN THE LOAD

In 1991 the Health and Safety Executive launched 'Lighten the Load', a three-year campaign to reduce musculoskeletal injuries at work – the largest workplace-caused incidence of absenteeism. The campaign concentrated on:

- work-related upper limb disorders (WRULDS), more commonly called repetitive strain injuries (RSI)
- the introduction of new regulations for manual handling and display screen equipment in response to the European Directives

- ergonomic design (fitting the machinery and tasks to the human body's capabilities) to prevent needless injuries.

Correcting these types of problems involves:

- modifying how a job is performed
- changing the position of the seat
- relocating parts of the machine.

A 1990 Labour Force Survey reported that musculoskeletal disorders account for more than half a million cases of work-related illness per year. Two separate categories of workers are at high risk for these disabilities – those doing very heavy physical jobs and those in very sedentary jobs. Surprisingly, the difference in rates of back pain between the two groups is small. Musculoskeletal problems have become increasingly widespread as the use of new technology enters the workplace. These problems involve muscles, tendons, joints and bones and usually develop in the joints of the neck, back and limbs, particularly the hands and arms. Symptoms include pain, numbness, restricted movements and inability to perform tasks requiring specific motions.

You can be injured in many types of workplace but nursing, office, house and hospital cleaning, construction, laundry and industrial work, poultry processing, clothing manufacturing, computer keyboarding and assembly line work are among the most hazardous. Many of these jobs, which are mainly filled by women, require a great deal of heavy physical labour. Manual handling (lifting, carrying, pushing or pulling heavy loads) is connected to about a quarter of all accidents reported each year – back injuries and sprains and strains to other parts of the body.

> I buffed car seats, trimmed them and threw them on the line. We lifted up heavy frames, 16-pound frames, throwing them on the line, turning them over and turning them over again. I was standing the whole eight hours bending over, lifting up the car seats, frames and cushions.
>
> Carla, US car seat fitter

Many workers engage in jobs that the average person does not connect with heavy lifting. For example, domestics have to move furniture, buckets and heavy cleaning machines. Office workers in charge of records have to move boxes of records and files while kitchen staff lift heavy pots and sacks of food.

Whatever the job, the risk factors from lifting and pushing heavy loads are likely to be the same. These include inadequate mechanical aids, inadequate staffing, unsuitable workplaces and methods of work, overly heavy loads and work rates, inadequate educational and training procedures and lack of clear cut safety guidelines. Furthermore, lifting or pushing heavy weights is one aspect of work which puts an extra strain on the pregnant worker, and can lead to miscarriage.

Workers are now covered by the UK's Manual Handling Operations Regulations fulfilling the requirements of the European Directive on Manual Handling which went into effect on 31 December 1992. The regulations require employers to:

- avoid the need for risky manual handling so far as reasonably practicable
- assess all risky manual handling operations
- reduce the risk of injury to the lowest level possible
- provide employees with information about the loads.

While pregnant workers must take care not to exert themselves, in general they can continue to perform familiar occupational tasks requiring a good deal of strenuous activity. This is frequently less stressful than being switched to a new but physically less demanding position which can be more trying emotionally. With some modification, this also holds true for jobs which require lifting and pushing heavy weights. The general rule of thumb is that if you can handle the load easily when not pregnant, you probably will not be unduly stressed during your pregnancy except for in the last few months, when it is probably advisable to reduce the maximum load lifted by 20–25 per cent. Regulatory bodies have generally discarded restrictions on the amount of weight to be lifted by women whether or not they are pregnant – that limit depends on the physical strength of the woman and how she feels during pregnancy. The strain from pushing or lifting heavy weights varies from woman to woman and can differ from one pregnancy to another.

It will take time for all workplaces to meet the requirements of the new legislation. In the meantime, proper lifting techniques will minimise injuries to your body or your pregnancy. Avoid lifting in front of your body during the last trimester as it places a burden on the lower (lumbar) region of the spine which is already under stress as a result of your weight gain, the expansion of your uterus and your loosened joints. You should be aware that two adrenal hormones, epinephrine and norepinephrine, increase during strenuous activity. Epinephrine quiets uterine muscles but norepinephrine increases the muscular activity, possibly inducing premature labour. If mechanical lifting aids are not available, either divide your burden into smaller units or ask for help.

Manual handling and lifting where heavy loads and inadequate lifting devices are the culprits and where jobs are poorly designed so that workers have to reach or stretch awkwardly are nothing new, but these chronic problems have been exacerbated by the introduction of technology which increases the number of jobs involving highly repetitive work that has to be completed rapidly without accompanying safeguards for worker health.

Repetitive Strain Injury (RSI)

RSI in most industrial countries is already, or soon will become, the largest cause of workplace injury. The TUC estimates that RSI costs the British economy a billion pounds a year and in 1994 a management consultancy firm reported that interviews with over 3000 British VDU operators revealed that 10 per cent exhibited numbness, tingling fingers and muscular pain of the upper arms and symptoms associated with RSI, and a further 1 per cent had already been diagnosed as having RSI. In addition, one in five operators who spent at least 75 per cent of their time at a VDU had RSI symptoms. Many of the sufferers had adequate workstations and environments. The consultancy firm concluded that the human body was not suited to all day computer work and that jobs would have to be redefined in order to avoid injury.

RSI in the Office

Most studies of office workers agree that the amount of time spent at a computer keyboard each day is the main risk factor in the development of RSI. However, in a 1991 European Union survey other factors were found to affect the development of RSI as well, including the amount of influence the worker had over his or her work and work flow. Women were found to have less influence over both these aspects than their male colleagues. They have less autonomy than their male peers and do short repetitive tasks more frequently. In fact, 27 per cent of women workers in the EU have jobs which consist primarily of repetitive tasks leading to monotony, stress and RSI.

The publicity that the female keyboard operators received after they won damages for Repetitive Strain Injury from British Telecom (BT) in 1991 did more to alert women about the possible hazards they face and the permanent damage that could result than anything else. Not only did this case alert women, however, but it also alerted employers to the fact that it might cost them more money to maintain their present unhealthy practices than to provide a healthy work environment.

In the late 1970s and early 1980s, many employers were woefully ignorant and wilfully uncaring about possible injury to women who keyed in data to produce invoices and accounts for things such as telephone bills. BT employees were ordered to type 13,000 characters an hour, which came to 3.6 letters a second. If they did not reach that target, their machine automatically registered a dock in pay and a report to management. If such practices indicate progress, the factory girl on four cents an hour at the beginning of the book may not have had it so bad!

(Further detailed information about VDU work in the office can be found earlier in the chapter in the subsection VDUs: Stress on the Work Front under the main heading The Office: Hidden Dangers.)

RSI in the Pottery Industry

Both the poultry and pottery industries have also garnered a good deal of publicity. They both have a history of large numbers of cases of Repetitive Strain Injury (RSI); it is a serious occupational hazard, for example, for the women who work in the Staffordshire pottery industry. Most of us think of cups as a drinking utensil, but to the women who make pottery cups what comes to mind is the approximately 6000 items they manipulate during an eight-hour shift. A handle fettler removes the seam from cup handles before sponging them smooth, manipulating small items of clay thousands of times a day. Cup spongers also handle small items of clay. At a piecework rate, they have to work very fast in order to make a living and the amount of clay they lift day after day adds up.

The management, faced with the loss of valuable skilled workers and mounting compensation claims, implemented a preventive health policy which included altering the damaging work process, moving workers with symptoms to alternative tasks, allowing for lower productivity and designing special splints that cut out movement at the wrist joint but enabled workers to move their fingers, elbows and shoulders. These splints help distribute the range of movements usually borne by the wrist over a wider range of joints and, hopefully, prevent deterioration and recurrence of RSI. They are not a cure, and some doctors even claim that they do not aid healing, but they do allow some affected women who are their families' primary support to keep jobs that they badly need.

If you feel any symptoms of RSI – numbness, pain or tingling in your fingers and arms – report it at once, especially if you are pregnant. You may be offered a job which doesn't involve handling quite so many items or where the pace of work is slower. The splint makes work a bit more awkward and if you are in the late stages of pregnancy, the splints will further restrict your agility.

What Can Be Done About RSI?

Do not underestimate the amount of permanent damage that RSI can cause. For some – but not all – injuries involving muscles and ligaments, a full recovery can be expected. The main treatment is rest. Anti-inflammatory drugs are sometimes needed and in very bad cases surgery is required: even then some suffering permanent injury may have to give up their jobs. Some women who have suffered from severe cases of RSI also have trouble carrying out their domestic and parenting jobs. The best advice is to try to prevent damage in the first place. Unfortunately, after you seem to have healed, the injury will recur if you go back to a job requiring the same motions that initially caused it – sometimes this happens even if the equipment has been redesigned according to proper ergonomic standards. If you have had a bad case of RSI, you may have to request a change to another type of job within the firm.

Many doctors do not immediately associate the workplace with some of the repetitive strain symptoms that their patients complain about; it can be misdiagnosed as workers suffer from a wide range of conditions characterised by persistent pain or discomfort without clearly visible injury. There is much disagreement in the medical community about the diagnosis and treatment of RSI, which makes it more difficult for sufferers to get appropriate, early treatment.

The HSE, aware of the need to sensitise doctors to this problem, published a leaflet for family doctors highlighting causes, symptoms and treatment of work-related musculoskeletal problems. In addition, as part of its 'Lighten the Load' campaign, it has produced several free leaflets and booklets. One for employers includes advice on the importance of being aware and concerned about musculoskeletal injuries caused by improper workplace design and processes, and on how to assess the risk and how to reduce it. One for employees explains how to recognise symptoms that might be early warning signals for workplace-induced injuries and who to inform about these symptoms – your employer and your doctor and, if you have them, a works nurse or doctor and union representative. Your employer has a legal duty to safeguard your health and safety and both you and your employer should work together to identify tasks which could cause problems and take steps to correct them.

The TUC has also been very concerned about RSI and in 1993 launched a national RSI campaign called 'Don't Suffer in Silence – Help Stamp out Upper Limb Disorder at Work'. In fact so many workers suffer from RSI that a British RSI association was founded. In 1990, it carried out a survey of its members. Of the 186 responding, more than three-quarters were women and about half were trade union members. Nearly two-thirds had suffered from RSI for three years or more and more than half of those taking part in the survey had lost income due to RSI. Only a quarter of employers had responded to the reporting of symptoms by improving the workplace. Employers will, however, have to introduce ergonomically sound workstations in the future in compliance with the Health and Safety (Display Screen Equipment) Regulations 1992.

For a few specified conditions, you are entitled to state compensation under the Social Security (Industrial Injuries) (Prescribed Diseases) Regulations 1985. Ask your doctor whether you might be eligible or request leaflet NI2 from your nearest social security office. If your workplace wants to start its own training and preventive education effort, they should contact the HSE's Information Line (0541 545500). If you need confidential advice, you can also call the Employment Medical Advisory Service at your local office of the Health and Safety Executive for further specific details. (Local addresses for the HSE are given in the Useful Addresses section at the end of the book.)

NOISE AND VIBRATION: CAN THE FOETUS BE HARMED?

Almost 2 million UK workers are considered to be at risk of harm from excessive noise and about 30 per cent of industrial workers are thought to have some work-related hearing loss. This may be due to other causes besides noise or a combination of noise and other exposures. New research indicates that exposure to heat and five organic solvents (carbon disulphide, toluene, styrene, xylene and trichloroethylene) in conjunction with noise may affect your hearing and balance. Unfortunately, damage may even occur when noise levels are at currently acceptable levels.

Assembly line workers, airline attendants, garment and textile workers and construction workers among others are exposed to a great deal of continuous noise and vibrations as part of their jobs. The revving of aeroplane engines, the continuous clanking of the assembly line, the whirring of sewing machines and the roar of construction machinery take a toll on workers' health. Noise affects more than hearing and can cause more than headaches. It can make your blood pressure rise and long-term exposure may be a contributing cause of heart disease. By blocking out warning signals, noise can also constitute a safety hazard.

But can a high level of noise and vibrations also affect pregnancy and injure a foetus? We know from studies and personal anecdotes that the foetus responds to noise and vibrations in the mother's environment. One study found easily distinguishable foetal movements when an automobile horn honked a few feet from the pregnant woman. In several cases, mothers reported foetal movements immediately after sound stimulations. A pregnant miner who worked in an area where there was noise from sledge hammers and cars carrying coal reported that her foetus was more active when she worked in the mine and quietened down after she left work. Whether this receptivity to noise external to the womb causes it harm is not definitively known, however, results of animal studies do raise suspicions as to the safety of occupational noise and vibration regarding reproduction. Some of the pregnancy effects seen in animals are spontaneous abortion, premature delivery and toxaemia. One hypothesis is that exposure to excessive amounts of noise and vibration disturbs the circulation of the mother's blood in the uterus.

The combination of information gathered from animal studies and human adults suggests that it is wise to avoid undue noise and continuous vibrations while pregnant until such time that conclusive evidence is gathered as to the foetus's vulnerability. But at a 1992 international conference preliminary reports indicated that noise levels typical of most jobs – those free of unusually loud noises – probably pose no risk to the foetus. Because foetuses can hear during the third trimester, scientists are concerned about the possible effects of high levels of sounds and vibrations on the foetus during this period. As noise is a stressor, the effect of high levels of sound and vibration on the pregnant women is also of concern.

In many worksites employing women, it is relatively easy to relieve the problem of excessive noise and vibration. As a result of a few relatively simple changes, all workers – not only those who are pregnant – will benefit from this reduction. Machines can be designed to ensure minimal noise and vibration and proper maintenance can help reduce the noise of existing machinery. Noisy processes can be isolated in one section of the workplace using noise absorbing material in floors, partitions, walls, curtains and ceilings. By implementing some of these measures and providing frequent rest periods and job rotations, your employer can decrease your exposure.

The 1989 Noise at Work Regulations require that employers reduce the risk of hearing damage to the lowest level reasonably practicable. The catch is that these regulations, based on a European directive, omit the directive's requirement for technical and work organisation measures to reduce exposure and interpret the 'reasonably practicable' phrase very liberally. As a result, the main effect of the regulations has been to increase worker use of ear defenders rather than the elimination of the noise itself. This is not a particularly useful solution as it has been found that removing hearing protection for as little as 15 minutes during an eight-hour work shift can cut protection by as much as a half. Moreover, ear defenders must fit properly, and in general, workers do not always receive correctly sized protective equipment.

During the European Year of Safety, Hygiene and Health Priorities which ran from 1 March 1992 to 28 February 1993, noise and vibration was one of the areas chosen for particular attention. The HSE will be able to provide you with references to research and programmes about ameliorating workplace noise and vibrations (HSE information line: 0541 545500).

SANITARY AND SAFETY PRECAUTIONS: DISEASE AND DANGER

Pregnant laundry workers, healthcare personnel, laboratory workers, airline flight staff and teachers (particularly in primary schools) often come into contact with people who have infections or material that is contaminated. Nora, a pregnant flight attendant who has worked in her job for thirteen years, recalls passengers collapsing from hepatitis and children taken off the plane with chicken pox. She says that she finds herself looking at passengers to see if they look healthy and, if they do not, trying to guess what illnesses they might have.

Although the pregnant woman does not seem to be more susceptible to infection, once she does become ill, the infection may be more severe and require a longer recuperation period. Certain diseases such as rubella (German measles) can cross the placenta and harm the foetus, as can some drugs given the mother to combat an illness. These can result in an early miscarriage, foetal infection, abnormality or death.

The following prescription drugs should be avoided during pregnancy if at all possible:

- dilantin
- cortisol
- androgens
- chemotherapy (anti-folate agents, alkylating agents)
- diethylstilbestrol
- tetracycline
- opiates
- benzodiazepines.

Also take extra precautions if your job involves manufacturing or administering any of these drugs. If you are in doubt about taking a specific drug, check with your GP or a pharmacist at a chemist. If you are worried about manufacturing or administrating a specific drug, check with your safety representative.

Another hazard is improperly treated rubbish and disposable waste in hospitals, manufacturing plants and office buildings. For example, a chemical which should be disposed of in a leak-proof container is spilled into a sink instead. In the sink drain it combines with the remains of another chemical that should not have been poured in the sink either. The newly combined chemical then bubbles up into a sink in a different room subjecting other workers to a hazardous substance of which they are totally unaware.

In addition, the workplace itself may be dirty – spills are not immediately wiped up, cockroaches and other insects cohabit, clutter accumulates in the halls and workstations, floors and stairs are slippery and machinery is not maintained properly. You are especially vulnerable to existing safety hazards during the last trimester when your body is harder to manoeuvre and you have little agility.

SHIFT WORK: BODY RHYTHMS OUT OF CYCLE

Medical evidence indicates that working rotating shifts poses psychological and physical hardship because our circadian rhythms (body functions that vary systematically throughout the day, such as temperature and sex drive) often become out of phase with the rest of our activities. This affects digestion, the immune system, sleep, alertness, motor reflexes, motivation and powers of concentration. It is thought to be why the number of accidents reaches its peak between 4 a.m. and 6 a.m.

Shift workers on average smoke more heavily, are more obese, eat less nutritional foods, participate in fewer leisure activities and are less involved in social networks, have higher cholesterol and serum triglyceride levels – all risk factors for cardiovascular disease. Women who work shifts show high levels of job stress, emotional problems and more frequently use sleeping pills, tranquillisers and are heavier drinkers than their counterparts who do not work shifts. Pregnancy causes physiological changes, shift work upsets natural body rhythms, and the combination can upset the body's functioning. Therefore, if you work shifts, you are more likely to suffer ill effects from your variable schedule after you become pregnant than you did before.

Working night or swing shifts (usually 3–11 p.m.) aggravates the fatigue of pregnancy and can reduce appetites to an unhealthily low level. Even if you are hungry, you may not be able to find any food at all because the cafeteria is closed or it is unsafe to leave the building at night. You lose touch with your family and friends and often cannot visit or attend social occasions as you need to sleep while others enjoy themselves. Women become more isolated and lose a valuable support network, increasing the stress. What is even worse for people than ordinary shiftwork is rotating shifts, as the pattern of rotation continually upsets the body's eating and sleeping cycles.

> We had to work overtime and on Saturday and Sunday if there was a lot of work. It was tough because we didn't have set hours every day. Having to get up at a different hour every day, eat at totally whacked out hours would put a strain on any normal person. You could work 9–5 one day, 11–7 the next day or 2–10 the next. They could have been more accommodating since there were five of us. But I think they had a bad attitude. Our supervisor took it as a personal thing that we all got pregnant at once.
>
> Maria, US lab technician

Many shift work jobs are particularly stressful and fatiguing because they are boring, repetitive and allow you no control over your work. Fatigue and lack of appetite are more than nuisance problems because they increase the likelihood of your giving birth to babies weighing less than five pounds. These low birth weight babies face a higher risk of dying within the first few months of their lives than babies who weigh more at birth.

The only time that a night or swing shift provides benefits is when it also provides additional autonomy. Pam, for example, worked the night shift in the sterilisation department of a hospital. She was able to switch from a sitting to standing position whenever she felt stiff and could work at her own pace because there was only one other worker on duty at night. During the day many employees and supervisors were on duty making sure that the rules and regulations were being followed.

For most people, however, shift work does not offer the freedom that Pam had, and in these cases it is important to try to minimise the strain on the body. If you are on a rotating shift, request that you do not rotate daily or weekly but stick to one shift for at least three weeks so your body can stay adjusted for a period of time. On some types of jobs this may not be possible, but do ask – unless you and your co-workers are assertive nothing is likely to be done!

SMOKING: PASSIVE SMOKING – A NEW REPRODUCTIVE THREAT?

As further research has been conducted, the health implications of passive smoking (the cigarette smoke in the air breathed in by non-smokers) are getting more frightening. A report issued in 1992 by the United States Environmental Protection Agency likened passive smoking to exposure to asbestos and radon. While the British health establishment has started an anti-smoking campaign, the percentage of smokers is still high and remains higher than in the United States where strong state and local regulations forbidding or limiting smoking in the workplace have been enacted. In many firms, employees who smoke either have to leave the building to smoke out of doors or are relegated to a few smoking locations.

An article in *Work, Environment and Health* concludes that although much remains to be known about the effects of passive smoking, there are already sufficient grounds to restrict or prohibit smoking in the workplace as passive smoking has been linked to an increased risk of developing lung cancer or emphysema. The Health Education Authority has launched a revised charter calling on employers to provide additional smoke-free zones for employees and to install a comprehensive smoking policy. In 1992, following consultations between unions, staff and management the Midland Bank instituted a comprehensive no-smoking policy throughout its 1855 branches, and other companies followed. Customers were also requested to observe the no-smoking rule. No-smoking policies are beginning to gain favour: a recent survey by the Consumers' Association found that 40 per cent of those polled wanted restaurants smoke free, 50 per cent want smoking banned in cinemas and 68 per cent support smoke-free shops, buses and trains.

> I do feel that my baby was at risk because of the smoke in the unit and the lack of ventilation. There are studies now that passive smoke is more dangerous than it was previously thought to be. There are problems with low birth weight babies in places where there is smoke, so I do feel that the baby is at risk. I really don't know yet because I haven't had the baby.
>
> Adie, US psychiatric nurse

> The smoking bothered me. It was a long struggle to get the staff to stop smoking in the staffroom, but now that is better.
>
> Hannah, part-time primary school teacher

> The ventilation is very poor and you have a tremendous number of smokers on the plane. You have to work in the smoking sections whether you want to or not.
>
> Nora, US flight attendant

Unfortunately, unlike Adie, Hannah and Nora too many women who try to eat a healthy diet during pregnancy, avoid drinking alcohol and performing dangerous tasks in the workplace still smoke while they are pregnant. Smoking while pregnant can cause your baby to be born with a low birth weight which is linked to poor growth, infant illness and death. A recent Danish study showed that even when the mother did not smoke, but other members of the family did, the baby's birth weight was affected. Therefore, smoking by the father can affect the health of the foetus as well. An increased risk of childhood asthma has also been linked to parental smoking habits and a National Child Development study following the progress of all Britons born in the third week in March 1958 until the age of 23 found that those whose mothers had smoked during pregnancy were shorter in height and less likely to achieve high grades in school than those whose mothers did not smoke.

Advice on how to stop smoking is now more likely to be part of routine antenatal care. Your midwife or GP can offer some useful techniques to help you stop smoking if you really want to or will inform you about support groups who can help. Some unions are interested in helping their members to avoid passive smoking in the workplace. Arm yourself with information, and disseminate this to people who smoke in your home, social and work environments. Most are now aware of the harm they are causing others and will try to be courteous about their smoking even if they do not stop themselves, although others may be abusive and inform you to mind your own business. But their smoking in a shared environment *is* your business and the stakes to a pregnant worker are too high for her to keep quiet.

STRESS: THE WOMAN CAUGHT IN THE MIDDLE

Stress is a well-recognised work hazard for executives, managers and professionals as well as lower-paid blue collar and service sector workers. Stress, however, is particularly severe in jobs characterised by high demand and low control – the type of jobs that are overwhelmingly occupied by women. Even when men fill the same positions, they are often given greater scope for decision making than women.

Excessive and chronic stress is to be avoided for everyone, but especially for the pregnant worker. As was discussed previously many of the body's reactions to stress, such as a rise in blood pressure, an increased heart rate, digestive disturbances and greater blood flow to the limbs, compound existing stressful – and sometimes dangerous – changes resulting from pregnancy. Recent research indicates that emotional stress constricts the blood supply to the uterus and placenta.

This is not to imply that pregnant workers cannot tolerate moderate, occasional amounts of stress. In fact, some stress is useful in re-evaluating pregnancy needs on the job. But chronic workplace stress can cause the pituitary gland to stimulate quantities of adrenalin more appropriate for emergency situations rather than everyday activities and this is unhealthy.

Reducing stress in the workplace is not always expensive. Sometimes small, well-planned changes in the workplace that cost little or no money can make a big difference.

> I worked evenings from 5.30 to 9.30, and at 7.30 we had a fifteen-minute break. We could have eaten in the cafeteria, but it was impossible to reach the cafeteria and eat in fifteen minutes. If the employer could have given us an extra five minutes for our break we might have been able to eat dinner in that time.
>
> Darcy, US hotel housekeeper

Even if Darcy's employer had refused to give her the extra five minutes, she could have asked to start her shift five or ten minutes earlier in return for a slightly longer break that would allow her to eat dinner and obtain adequate nutrients for herself and her foetus.

While we know that stress makes us feel uncomfortable and sometimes ill, we do not know whether it will harm our unborn children. Unfortunately, scientists in many cases can indicate the possibility of harm but cannot provide any firm conclusions.

The following lists are of primarily women's occupations which have high stress levels, and of common causes of stress in jobs such as these:

Highest stress level

- Waitress
- Bank teller
- Cashier
- Nurses aide

- Assembler
- File clerk
- Receptionist
- Sewing machine operator
- Telephone operator
- Office machine and keypunch operator

High stress level

- Cook
- Sales clerk
- Typist
- Secretary
- Library clerk

Common work-related causes of stress

- No control over speed of work
- No control over variety of work
- Repetition of monotonous work
- Not able to leave workstation
- No control over decision making
- Insufficient time to complete assigned work
- Insufficient opportunity to obtain a promotion

In addition to stresses at work, the pregnant worker often finds herself in a particularly stressful situation trying to combine the roles of worker, mother and/or mother-to-be and homemaker. Supermum is not a figment of journalists' imaginations but an accurate portrayal of the diverse tasks women juggle in their daily lives. The term is particularly apt for the pregnant worker with other children at home.

Physicians see substantial numbers of patients suffering from stress and now consider it a factor in a vast majority of physical and mental ailments. The list below shows what a wide variety of health problems may be stress related. Birth defects are included as possibly having a stress component. This possible link between stress experienced by the pregnant woman and a baby being born with a birth defect is a worrying one. Some writers have even suggested that stress caused by a variety of factors including interpersonal tensions may affect the development of the foetus leading to neurological difficulties, physical defects, slow development and behaviour disturbances in the child.

Health Problems Thought to be Associated with Stress at Work

accidents	birth defects	indigestion
alcoholism	colitis	infections
anxiety	depression	insomnia
arthritis	fatigue	neck pain
asthma	headaches	skin rashes
backaches	heart attacks	strokes
battering	high blood pressure	ulcers

Work-related stress can arise from physical conditions of the workplace or from interpersonal sources. Our jobs can be monotonous and monitored, our co-workers unpleasant and uncooperative, our superiors unfair and unsympathetic, and management policies can be inequitable and inflexible.

> My work is very stressful. Everything has to be done *now*! But I can cope with it. When it is busy, I work from 9 a.m. to 8 p.m. I felt very tired at the beginning. I delegated more work when I was tired.
>
> Catherine, chartered accountant

> For the last few weeks the stress of the job is hard. I've been more tired than usual and I have a little one at home. I now work part time since I had my son who is eighteen months old now.
>
> Heather, hospital clinic administrator

Despite the evidence that sources of stress stem in the main from the physical and interpersonal structure of the workplace, many doctors, books and courses advise the woman worker who is planning to become pregnant, or who is already pregnant, to handle stress by individual techniques such as relaxation exercises, yoga and meditation. These are undoubtedly helpful, but why is the burden put on women to counteract the stress rather than the emphasis being on reducing it at the source?

> I ran the whole kitchen on a day-to-day basis. I made salads, chopped foods. Cutting puts strain on the abdomen and back and pushing trolleys to set up buffets pulls on the body. In the summer the kitchen was too hot, the fans were inadequate. I had high blood pressure during pregnancy. The doctors did not know what caused

> the high blood pressure, but one possibility was job stress. There are deadlines every day. Regardless of whether the staff comes in or the food arrives, the show must go on.
>
> I would come in and vomit every day for months and I would faint. I couldn't think food. It was hard to plan menus. I put tablecloths on the floor in my office and made a makeshift bed. If I felt faint, I would lie down. If I didn't have the extra bits, it would have been awful.
>
> Mary, catering supervisor

The mental health charity MIND cites Department of Health figures indicating that about 80 million working days per year are lost through stress-related illness, while only about 700,000 days are lost through strikes. In the UK, about 10 million people a year are thought to be under such pressure at work that they perform inefficiently, are accident prone and feel that they have to take sick leave to have a break from the pressure. According to MIND, women are more likely to have stress provoking experiences at work related to either sexual harassment or coping with family responsibilities while holding down a job. Many working women complain that their employers are unsympathetic to either of these problems.

Social Stress

In addition to physical stress, social stress poses problems for pregnant workers. These usually stem from a clash between your own needs and those of the management. As a result, you can face social stress from several different aspects of your job. Inflexible work rules and shift work compound the normal fatigue of pregnancy. Friction with your supervisors and co-workers can make your work lives miserable, while unclear guidelines and childcare difficulties can leave you continually worried.

> The worst people who are not sympathetic are the women with no children. My boss decided not to have children, and the workers were afraid to tell her they were pregnant. When they found out I was pregnant, they said, 'You're not going to tell her. She'll be so angry.' She needed me so badly that she didn't cause trouble. She didn't say

anything. It was the look. I had a doctor's note about my high blood pressure so she had no choice.

If I hadn't been pregnant, I wouldn't have been so understanding to my staff. I always watch out for hazards, tell pregnant women not to run. I try to look after a pregnant person to make work more comfortable. I change their tasks. They have a stool so they can sit for a while. Pregnancy has really changed my outlook. I became calmer and if I could have got through the pregnancy, I can get through anything now.

Mary, catering supervisor

Friction With Supervisors and Co-workers

Women report how upsetting they find disputes with their supervisors about their work and pregnancies. Some researchers think that severe interpersonal tensions may affect the development of the foetus, leading to physical defects, neurological impairment, slow development and behaviour disorders in the child. Luckily, most pregnant workers find that their supervisors are understanding and co-workers become protective of their health. When the pregnancy is resented by others, however, your life can be made miserable. Instead of being supported, you are harassed.

The supervisor/supervisee relationship is a delicate one even when you are not pregnant, and a poor relationship can make your workday miserable.

The new principal was a real humdinger. She's a drill sergeant type of person and she particularly is interested in having the place clean and neat. At eight months pregnant, I was on top of the chair dusting off the top of a coat closet and I thought to myself, 'this is stupid'. Then I thought that all the emotional stress that she put on me was bad for my baby. I feel she caused my high blood pressure problem. This woman obviously resented me being pregnant and on her staff.

Dinah, US primary school teacher

The Women in Work and Family Life study, conducted in the 1980s at Bank Street College in New York City, cited improvement in the boss–employee relationship as being the most important factor in women's jobs that would spill over to improving their family life as well.

> My boss wasn't very happy when she found out I was pregnant. She told me right up front that she hated the idea of my being pregnant and wanted to know what I was going to do about it. I did everything I did when I wasn't pregnant and then some because I had to show her that I could still do my work and function the way I would if I was not pregnant. I guess she wanted to break me or break my spirits so she always had something extra and I never knew what that was until I'd come in.
>
> Pamela, US office worker in a residential treatment centre

> The union issues literature, but I do not know of any specific policy toward change in work or workload during pregnancy. It depends upon the manager. My manager who is a man has been really good. He tells me to stop worrying about rushing around to get things done and sometimes helps with the filing if I have to bend down.
>
> Sorya, clerical assistant

> My boss has an attitude problem. He told me that I chose the wrong time to get pregnant. He wanted to know who he could get to do my work and said that he'll have to do it. I felt very hurt and guilty and he made me feel that I couldn't be pregnant. Every once in a while he goes on about it. He made it seem as if I put a great burden on him. I was made to feel like I was a machine. I was told that the boss is a misogynist. He thinks a woman's place is in the home.
>
> The personnel officer who had no children of her own kept telling me what to do but she wasn't helpful. She wanted a certificate from the doctor each week saying that I could work. She was very stressful. The personnel officer smiles all the time, but she is a cow. My boss also sticks to the book. He was surprised that I was coming back before my statutory leave was up. I was told that he wanted to save money and he couldn't save as much if I came back early.
>
> Amira, librarian

Co-workers can be as significant as bosses in contributing to the stress level of the workplace. Being in continual contact with people who dislike you is extremely draining. Caroline, for example, a telephone operator for a doctors' answering service, was harassed by the other operators who showed no concern for her pregnancy. They suggested that she leave rather than reduce their smoking.

> The women urge you to quit. They would urge anyone to quit because they've been there for years and it's like the new kid on the block. You know you get the hard time. So they're no help. They don't encourage you. We don't work together the way we should. There's too much tension.
>
> Caroline, US telephone operator

Thus the attitude of bosses and co-workers can make or break a job, especially if you are already coping with nausea, fatigue, swollen ankles or backaches.

> I didn't feel that other staff members were supportive during my pregnancy. In other places, people seemed to help pick up and swap off duties when someone wasn't feeling well and I didn't find that the case here. I think there was a difference between the male and the female attitudes. The males seemed to be very protective and wouldn't let me get involved in anything that might be detrimental to my health. They wouldn't let me lift, they wouldn't even let me sit one-to-one with a patient who was having any kind of difficulty at all. The women felt that I could do anything that anyone else could do.
>
> Adie, US psychiatric nurse

Female co-workers may feel hostile or angry toward pregnant colleagues for many reasons; because they fear that they will create more work for them or perhaps because they are jealous of women who combine motherhood and work when they feel that they have had to choose between one or the other. As more and more women work outside the home during their pregnancies, sceptics will learn that pregnancy is not a barrier to effective job performance. You may even serve as a role model for other women who are uneasy about attempting the joint venture of paid employment and pregnancy.

If you make sure you pull your weight and do not put a burden on others to do more than their fair share, you will minimise discord. At the times when you do perhaps need special arrangements during your pregnancy, explain to your co-workers and supervisors:

- what the problem is
- for how long you will need the arrangements to continue
- how they will or will not impinge on the other workers
- how you might be able to compensate for this special treatment.

If you are completing your work and no one else is being imposed upon, formal or informal agreements usually can be worked out.

Sexual Harrassment

There is one more area of concern regarding supervisors and co-workers – sexual harassment. Sometimes this does not stop during pregnancy and insensitive men continue to make lewd remarks and continue unwanted touching. Sexual harassment is serious – it can affect the victim's emotional and physical well-being long after the specific incident has ended. In fact the Health and Safety Executive includes sexual harassment under its definition of violence in the workplace.

If you are being sexually harassed:

- ask the harasser to stop and make it very clear that the actions are unwelcome; harassers often have thick skin and believe that the woman is just being coy and does not mean what she says
- if the unsavoury behaviour continues, keep a record of incidents, their time and date, circumstance and the names of any witnesses
- even if you decide not to pursue the matter, report the incidents to your union. It is important that the union does not underestimate the problem. It can negotiate a stronger sexual harassment policy if conditions warrant one.

Inflexible Work Rules: Give Us A Break!

You may find that inflexible work rules create more of a hardship when you are pregnant than the difficulty of your work: few jobs demand continuous physical activity; rather they require a large amount of exertion only part of the time. When you become pregnant, you need additional flexibility and you feel the restraint of rigid work rules even more. You need more frequent rest breaks in which you can walk around, change positions, stretch, eat a nutritious snack or use the toilet, rather than a reduction in your work load. Access to toilet facilities when you need them is extremely important because the uterus exerts pressure on the bladder, especially in the last few months. Increased frequency of breaks and flexibility in the timing of breaks allows you to continue performing your job efficiently.

> I was only allowed to use the lavatory on those planned recess periods, or at lunchtime which was inadequate for my needs. Fortunately, the children's bathroom with the low toilets was just across the hall. I was always running across little kindergarteners while I was in there. It was against the rules, but it was really necessary.
>
> Kathy, US primary school teacher

Sometimes facilities are inadequate even when the work rules are more flexible. For example, there may be no adequate lounge area in which you can rest or put your feet up, or no cafeteria where you can obtain nutritious food.

> We'd have nowhere to go during break. All we had was a bathroom which has two toilets. Sometimes when I was really tired, I'd go in there and sit on a toilet and other times I'd sit on a window sill. If anything, my breaks decreased because my boss wasn't very happy when she found out that I was pregnant.
>
> Pamela, US office worker at a residential treatment facility

VENTILATION: WHAT KIND OF AIR ARE YOU BREATHING?

Unless a building is properly ventilated, pollutants can build up to levels that are unhealthy and possibly dangerous to you and your unborn child. This is the situation found in many workplaces where indoor air pollution is a severe problem. The construction of energy saving plants, factories, hospitals, schools and office buildings with windows that do not open has led to this grave situation. Cigarette smoking, carbon-monoxide emissions from loading docks and garages, fumes from office and factory machinery and the increased number of chemicals used in work processes, furniture and equipment make the matter worse.

Because we hyperventilate during pregnancy in order to obtain the oxygen we need, we also inhale more pollutants along with the oxygen. Exposure to several toxic substances such as carbon monoxide, lead, benzene, and hydrogen cyanide can interfere with the oxygen-carrying capacity of the blood. In addition, they can cause cancer, respiratory problems, nervous disorders and sterility. Limit your physical effort in workplaces where these chemicals are in the air. Inhaling air containing these chemicals may cause you to experience only minor symptoms, so that you are not alerted to the more serious foetal effects.

> There was solder smoke. They said that was not really good for you by itself and even worse if you were pregnant.
>
> Marcy, US solderer and wirer

Working in an environment that is either too hot or too cold can also be very stressful.

> Ventilation is a problem for everybody whether or not you're pregnant. The air conditioning is hopeless. It just spreads bugs around. It is very intermittent some days hot and others cold and others the air doesn't move at all. Sometimes you get a terrible smell out of it, I think this is when they clean it. If the sun is out, it is too hot because the system can't cope.
>
> Heather, hospital clinic administrator

> It was very cold. The air conditioning was too cold all the time. I made many complaints, but they couldn't fix it. It was too cold because of where the department was located within the store. You could turn the air conditioning off, but then it got too hot.
>
> Louise, department store sales assistant

The key to cleaner air in your workplace is the type and maintenance of the ventilation system installed in the building. These ventilation systems should be designed to supply and circulate fresh air. The circulation of fresh air by itself is not necessarily a satisfactory solution as the outdoor air may also contain pollutants. Unless the system is correctly designed and in very good working order, it will not be able to eliminate even moderate amounts of toxins from either fresh or recycled air.

The blower which moves the air, the ducts which deliver the air and the vents which distribute it are the basic components of the ventilation system. The blower can be underpowered for the amount of space it is supposed to service and there may be too few ducts and vents or they may be dirty or blocked.

The following is a ventilation check list to help you to find out more about indoor air pollution. Check your workplace. You probably can come up with some relatively inexpensive and easy to implement suggestions that can be used as interim measures until the basic underlying conditions can be improved. For example:

- more frequent maintenance checks of the ventilating system
- changing from a toxic cleaning agent to a non-toxic one
- moving boxes or furniture located in front of the ducts.

As was previously discussed in the sections on temperature and lighting, hazards from these aspects of the workplace and from improper ventilation in the workplace are covered by the Workplace (Health, Safety and Welfare) Regulations 1992.

Ventilation Check List

- See if your workplace has a ventilation system. You can do this by looking for ducts and vents.
- Is the ventilation system on 24 hours a day if large duplicating and printing jobs are done at night? These machines can produce a high volume of pollutants. Hold a tissue near the vent after normal working hours. If it moves, then air is circulating.
- Does the ventilation system go on and off during the day? Some systems are on a time cycle regulated by a computer. This type of system gives inadequate amounts of clean air if pollutants are generated continually. Use the tissue test described above several times a day for a few days at different hours.
- Does each room have a supply vent and an exhaust vent? Again use the tissue test and see whether air is entering.
- Are the exhaust and supply vents right next to each other or blocked? If the two types of vents are too close to each other, the clean air gets sucked out of the room before it has a chance to circulate. If the vents are blocked by walls, partitions, boxes or files, the air flow will be obstructed and pollutants will not be eliminated efficiently.

- Are there 'dead spaces' in your workspace where no air is being replaced? Light a match and if the smoke does not move you can be pretty sure that the pollutants do not move either and just build up.
- Do work areas with copying and printing machines have adequate air supply and exhaust? For some machines extra vents near the source of the fumes are needed.
- Can workers control their ventilation by turning the blower or fan supplying the air up or down? Check with your building maintenance office to see whether this is possible.
- Is there a smoke detector in your ventilation system? For early detection of a fire, it should be located in the exhaust vent.
- Are the temperature and humidity adequate? Humidity makes a cold room feel colder and a hot room feel hotter. If the air is too dry, you may become more susceptible to colds and infections.

WORKPLACE EQUIPMENT AND DESIGN

Ergonomics is the science that attempts to adapt working conditions to suit the worker. Ergonomists, often called human engineers, design adjustments in the work environment to improve safety and efficiency while protecting the health of the worker. In the case of the pregnant worker, ergonomically sound equipment helps protect not only your body but that of your unborn child as well. Ergonomically designed work equipment and workspaces, allowing the flexibility to meet individual needs, should be a priority demand for pregnant workers in all types of jobs – executive, white collar, blue collar and service.

Most equipment is designed for an average employee and your body during pregnancy certainly does not fit that description. In fact, there is no average in a work population consisting of males and females ages 17–70. Pregnancy is only one factor in a highly diverse work force by any criteria – size, shape, age and health.

The human engineer can design flexible equipment that will both increase your comfort and improve the consistency and efficiency of your work. Seats, desks, tables, tools and workbenches of adjustable heights should be standard equipment in offices and factories. These can substantially reduce pressure on people's joints and muscles which comes from sitting for hours in awkward positions. Keypunch operators or workers in the garment and pottery industries, for example, can suffer from tenosynovitis, a tendon inflammation resulting from repetitive finger and hand movements in the same position. Sitting over a sewing machine or at a wordprocessor for several consecutive hours also causes muscle strain in the shoulders and back. If the work requires continuous standing, this may be a problem for women who are pregnant, as it can lead to swelling in the legs.

Problems regarding continuous standing are frequently due to work policy rather than work equipment. Many jobs, especially in the women-intensive service industries, require long periods of standing – saleswomen are frequently are required to stand even when there are no customers, for example. Many would feel a lot better if they were permitted to alternate between standing and sitting.

Personal Protective Equipment

Cleaning up the workplace is the ultimate goal we strive for, but wearing well-fitting personal protective equipment (PPE) may be a necessary, if disagreeable, interim measure. Until recently women seldom had well-fitting protective equipment; a prime area of concern as PPEs which are not the correct size do not provide adequate protection and are often uncomfortable, which may lead to them not being worn. Ill-fitting face masks and shields leave gaps so that dust and fumes enter. Hand tools are often too heavy to handle and hold and too clumsy and difficult to grasp, leading to undue strain and accidents.

The problem of inadequate PPEs is even greater for the pregnant worker who is more vulnerable to hazards than her non-pregnant co-worker and likely to feel more uncomfortable due to her body changes. Her body shape will be even more different from the scaled down men's sizes that are so often sold as appropriate for women. As anyone can see, women are not simply proportionately smaller than men in all body dimensions, in fact, they are proportionately larger than men in three areas – chest depth, hip circumference and back curvature at the hip – and at every height–weight combination women have significantly smaller shoulders. A pair of coveralls could thus be too small across the hips and too large across the shoulders. As the body changes during pregnancy, the problem of adequate personal protective clothing worsens. The Personal Protective Equipment at Work

Regulations 1992 are aimed at correcting these problems. (See Appendix C at the end of the book for more details.)

Equipment also needs to be redesigned to fit women's lifting capabilities. Because women are built differently from men a lifting task exerts 15 per cent more stress on a woman's back than on a male of the same size or strength. This stress is even greater for the pregnant woman, especially in the last trimester when they are more prone to back injuries. Equipment geared to women's lifting capabilities would also be better suited for many men as continual lifting accounts for a very large number of severe and disabling back injuries – one of the most common causes of industrial absenteeism. Poorly designed equipment is frequently responsible for these injuries rather than any inherent body weaknesses. The Manual Handling Operations Regulations 1992 (see Appendix C) aim to eliminate injuries due to lifting.

Workspace

Moveable lights, well laid out workspace, footrests and work holders make it possible for you to adjust the equipment to fit your individual needs. This is of particular value to you as your body changes substantially over the nine-month period. Such flexibility allows your body to be less stressed before you become pregnant and to return to its healthy non-pregnant state more rapidly.

Good seating is another priority for the pregnant worker. The backrest of the chair should be low enough to support your lower back and pelvic area. The seat should be cushioned, flat and wide enough to allow you to be seated with both legs in a supported position. A foot stool is an added bonus as it takes pressure off the legs and improves circulation.

Workplace design also includes space, light, noise, and air quality factors. If inadequate, these can pose harm to you and your unborn children by causing injuries and accidents and through physical stress.

FURTHER INFORMATION

It is wise periodically to update the information you have about reproductive hazards in the workplace and new protective legislation.

The information may change from one pregnancy to the next. Good places to start your search would be your safety representative, the TUC book catalogue, the City Centre, the London Hazards Centre, information points of the HSE, EU publications and occupational health departments of universities. (Addresses are given in the Useful Addresses section at the end of the book.)

CHAPTER 3 APPENDIX 1: HOSPITAL HAZARDS AND WHAT YOU CAN DO ABOUT THEM.

1. **Injuries**. These are primarily back injuries and sprained and broken arms and legs. You can try to avoid these by asking for help in lifting heavy patients and loads. Also be extra careful to watch where you are walking, particularly when you are in a hurry, are harried or stressed. These are the times you are most likely to trip or slip.
2. **Radiation and high energy exposure**. These come from X-ray equipment, particularly portable X-ray machines, radioactive dyes and implants. Large doses may cause tissue damage and genetic damage. Check with your employer, whether a hospital, general practice surgery, consultant or dentist, to insure that a proper maintenance schedule is in effect and that safe handling techniques are included in a required orientation programme for all employees. If not, complain to your safety representative, union or HSE. You are covered by COSHH regulations.
3. **Clinical hazards**. In the line of duty, certain routine jobs have anything but routine consequences. Sterilising chemicals, drugs and anaesthetic gases can cause skin and respiratory irritation, liver, kidney and nervous system damage, cancer and reproductive dysfunction. You must be extremely careful and wear gloves, lab coats and masks when warranted.
4. **Infections**. These can vary in severity from the common cold to a staphylococcus infection to hepatitis-B. The last is a serious liver disease for which luckily there is now a vaccine that is 90 per cent effective. Be sure that you take extra special care of yourself when you are pregnant. Make sure that you get enough rest, eat nutritious food and avoid stress whenever possible. In this way you will be less likely to catch a workplace related disease.

CHAPTER 3 APPENDIX 2: KNOWN AND SUSPECTED REPRODUCTIVE HAZARDS IN INDUSTRY

Art and Jewellery
boron and boric acid
cadmium and compounds
lead and lead compounds

Auto manufacturing and repair
aromatic hydrocarbons (benzene, toluene, xylene)
carbon monoxide
chlorinated hydrocarbons (solvents)
epichlorohydrin
formaldehyde
glycol ethers
heat, extreme
lead and lead compounds
vinyl chloride (VC)

Chemicals
anaesthetic gases
aromatic hydrocarbons
arsenic and compounds
boron and boric acid
chlorinated hydrocarbons
dimethyl sulphate
epichlorohydrin
ethylene dibromide (EDB)
ethylene oxide (EtO)
mercury
pesticides
selenium

Clothing, textiles and leather
arsenic and compounds
boron and boric acid
carbon disulphide
dimethyl sulphate
dimethylformanide (DMF)
epichlorohydrin
ethylene dibromide (EDB)
ethylene oxide (EtO)
ethyleneimine
formaldehyde
vinyl chloride (VC)

Electrical manufacturing
boron and boric acid
cadmium and compounds
carbon disulphide
polybrominated biphenyls (PBBs)
polychlorinated biphenyls (PCBs)

Electronic and semiconductors
aromatic hydrocarbons (benzene, toluene, xylene)
arsenic and compounds
cadmium and compounds
chlorinated hydrocarbons
glycol ethers

Food
chlorinated hydrocarbons (solvents)
ethylene oxide (EtO)
heat, extreme
pesticides

General manufacturing
aromatic hydrocarbons
cadmium and compounds
chlorinated hydrocarbons
epichlorohydrin
formaldehyde
glycol ethers
lead and compounds
styrene
vinyl chloride (VC)

Glass and pottery
arsenic and compounds

boron and boric acid
heat, extreme
lead
manganese
non-ionising radiation

Hospitals and healthcare
anaesthetic gases
carbon disulphide
ethylene oxide (EtO)
formaldehyde
ionising radiation
mercury

Offices
indoor air pollution
pesticides
video display units
stress

Painting
aromatic hydrocarbons
boron and boric acid
lead

Paper
chlorinated hydrocarbons
ethyleneimine
non-ionising radiation
polybrominated biphenyls (PBBs)
polychlorinated biphenyls (PCBs)

Pharmaceuticals
chlorinated hydrocarbons
dimethyl sulphate
epichlorohydrin
oestrogens
ethylene dibromide (EDB)
manganese and compounds
mercury and compounds

Plastics
cadmium and compounds
dimethylfomanide (DMF)
epichlorohydrin
styrene
vinyl chloride (VC)

Refining
aromatic hydrocarbons
carbon disulphide
ethyleneimine
lead

Rubber
carbon disulphide
chloroprene
lead
manganese

Steel
boron and compounds
carbon monoxide
heat, extreme
manganese

Wood processing
arsenic and compounds
boron and boric acid
ethylene dibromide (EDB)
formaldehyde
mercury and compounds

Source: Adapted from *Safer Times* fact sheet, Philaposh (Philadelphia Project on Occupational Safety and Health), Philadelphia, PA.

4

Reducing Risks and Hazards: What Future Mothers Need to Know

This chapter presents some basic information about evaluating risks and hazards. It explains why it is extremely difficult to accurately determine the harmful health effects of exposures to substances in the workplace and explains how to make judgements based on preliminary data. This information can help you learn to make some sense of what you read and hear so that you can begin to understand why scientists give conflicting findings and may disagree about the interpretation of existing evidence. Even within the Health and Safety Executive and Employment Departments, there is disagreement as to what data to accept or what standards to promulgate. Despite the uncertainty involved, knowledge about how reproductive health studies are designed and the basis on which conclusions are drawn can be used by you and your partner if you are thinking about becoming pregnant or are pregnant and are worried about the safety of your workplace. Some important questions for you to get answers to are listed later in this chapter and at the end of the book you will find a form called 'Occupational Information to Obtain for Your Midwife or General Practitioner'. You can photocopy these pages and can easily carry them with you if you decide to investigate further any possible workplace, home, or neighbourhood reproductive hazard that worries you. The glossary at the end of the book provides additional detailed explanations of scientific terms used and may be helpful in this effort.

LIVING IN AN AGE OF UNCERTAINTY: EVALUATING RISKS AND HAZARDS

One of the things that we find hard to deal with is the uncertainty in our lives. In the workplace, we feel that our employer has a great deal of control over our lives while we have very little. We worry about harm caused by the workplace which may be beyond our control and harm caused by losing our jobs if we complain too much, which is under our control.

Working while pregnant adds an additional dimension to the task of obtaining a healthy workplace because of the concern about the effect of occupational exposures on the foetus's health as well as our own.

> Everyone didn't want me to jeopardise the baby. It depended on who was on duty as to what they asked me to do. They asked me to quit at seven months as I was getting too heavy to do any lifting.
>
> Zena, auxiliary nurse

> Many of my pregnant friends who were flight attendants worked the initial six months the airlines allowed, and some of them had children with handicaps. You always wondered if the handicaps came from their having stayed. You really have no way of knowing if the proportion was any greater than the proportion of people in the ground jobs. I have a very good friend whose child ended up with cerebral palsy. I had a few friends that lost their foetuses during those six months. So you don't know. Did the plane carry some dangerous chemicals in the cargo or was it due to airline conditions, mainly decompression?
>
> Nora, US flight attendant

At Risk or Not at Risk? That is the Question

How can you know which of your fears are well founded and which are not? How do you know which workplace conditions should be complained about loud and strong? If you are unionised you can present your concerns and reasons for them to your safety representative, but many of you may not belong to a union, in which case the job of protecting your own health, that of your foetus and your children's health becomes even more difficult. You need to be continuously aware of new information about risks and integrate them into your daily life at home and in your paid employment. Many women read voraciously about 'do's and don'ts', especially when they are pregnant or contemplating pregnancy, but often there are so many conflicting statements and uncertainties. What good is it if scientists learn more about occupational reproductive risks in the future when you are planning to become pregnant now?

Women workers want health and safety standards in the workplace to be based on the most conservative estimates of risk to provide them with maximum protection. They are well aware that in the 'history of harm' a level asserted to be harmless one year may be considered harmful a few years later.

TRYING TO PROTECT YOURSELF FROM OCCUPATIONAL REPRODUCTIVE HAZARDS

The best way to protect yourself and your partner from occupational reproductive hazards is by ensuring that the workplace is healthy for everyone. If you are unionised, your safety representative can help you work toward this goal. This protects you not only while you are pregnant, but also when you are not. Sometimes, however, the extent, or even the existence of a hazard is unknown as tens of thousands of important commercial chemicals have not been tested extensively for toxicity. What is frustrating is that even when a chemical is shown to be toxic, it is harder to pinpoint reproductive harm than it is to determine other health effects.

One difficulty is that a substance causing harm in one species, e.g. a mouse or a rabbit, may not cause harm in another because of differences in the way the substance is processed in the body and differences in the reproductive systems of various species. For example, thalidomide, a drug taken by women in the 1960s during early pregnancy as an anti-nausea drug, caused their babies to be born without normal arms or legs. Thalidomide damages the foetuses of rabbits and non-human primates such as baboons, but not those of rats and mice which are the animals used in most toxicity tests.

The time at which exposure occurs may also be crucial for reproductive damage. Some hazardous substances act on the development of the sperm (spermatogenesis) but not later, while others can damage developing ovaries but not mature ones. Furthermore, some toxic substances only damage one function of one particular cell while others are toxic to many sites within the human body. Another major problem is trying to establish whether there may be a safe upper limit of exposure to substances that can cause cancer, genetic damage or developmental defects in humans.

After the thalidomide episode, when mothers who had taken the drug during pregnancy gave birth to babies with shortened or missing limbs, Britain instituted a National Birth Defects Registry in which detailed parental and environmental information is collected. In 1964, the Office of Population Censuses and Surveys (OPS) started collecting data on congenital malformations detected within the first week of birth in England and Wales. One purpose was to be able quickly to detect any sudden increase in the occurrence of any particular malformation in order to prevent another thalidomide like disaster. The Birth Defects Registry can provide valuable leads, but because understanding of reproductive hazards is clouded by such uncertainty at every level of investigation, the answers we need are not easy to obtain.

TYPES OF RESEARCH STUDIES

Scientists use animals, bacteria and humans to study toxic substances.

Animal Studies: Rats and Mice in the Laboratory

Rodents have played a major role in the scientific fight against human disease, with a large number of animal experiments being conducted on mice or rats. The studies compare rats and mice in experimental groups with those in a control group. Animal studies have been conducted to determine whether exposure to specific chemicals causes cancer or causes reproductive harm, although many more cancer studies have also been carried out.

Scientists face two main problems in applying findings from animal experiments to humans. One is that susceptibility among animals or between animals and humans differs. Certain chemicals give rats cancer but do not seem to harm mice or vice-versa. Not only do different species react differently to a specific substance, but individual members of a given species also can have different reactions. So, will they harm humans? Nobody knows for sure on the basis of these animal studies. The second problem is determining how accurately findings based on high doses of toxic chemicals given to animals over a short period of time reflect the risk humans face from the lower exposure levels they experience in their environments over a longer period of time.

Bacterial Studies: Millions of Microbes Under the Microscope

Considering the large backlog of substances to be tested, plus the thousand new chemicals entering the market each year, some experts feel that animal testing is too time consuming, too slow, too costly and too uncertain to continue to play the key part in the regulatory process. Some see a major role for new tests using bacteria instead, although others remain cautious and believe that bacterial tests are not as accurate as animal tests.

Several of these new bacterial tests can be completed in a month and cost considerably less than many animal tests which can take from four to seven years to complete. By measuring how toxic chemicals affect genes, chromosomes and other cellular systems, scientists may be able to predict more exactly the substance's capability of causing cancer, mutations, birth defects or environmental damage.

Human Studies: The Final Proof

After the evidence from animal and bacterial studies is accumulated, scientists then have to determine whether humans are going to experience similar effects. To help answer these questions researchers conduct epidemiological studies on fertility of male and female workers and pregnancy outcome. In some investigations, they compare the reproductive health of groups of workers who were exposed to certain substances with those that were not, or compare work histories of pregnant workers who suffered ill effects with those that did not to see what might have caused the harm. These epidemiological studies are primarily of two kinds – *retrospective* and *prospective*. The retrospective study looks at a group of people who have suffered reproductive injuries and a comparable group of those who have not. The researchers look for an explanation in their work and personal history. The prospective study follows the work and personal lives of a matched group of people into the future and investigates how those who suffer reproductive injury differ from those who do not.

For those of you who are interested in learning more about the scientific aspects of these studies, additional material is presented in the appendix at the end of this chapter.

A LINK BETWEEN CANCER AND REPRODUCTIVE HEALTH?

Scientists worry about workplace substances that might cause cancer as well as harming reproductive functioning. Will reducing workplace exposure to one type of risk also help eliminate the other? Very little is known about substances in the workplace that cause changes in human genetic material (mutagens) or affect the development of the foetus (teratogens). Researchers believe that there is a large overlap in the substances that cause mutations and those that cause cancer. Most carcinogens (agents that cause cancer) are mutagens (agents that change genetic material) and some mutagens are also carcinogens. If an agent is a known carcinogen, it also may be a reproductive hazard. Because of this overlap, some scientists feel that until improved techniques are developed for detecting mutations directly, one way to eliminate them is by having a strong policy protecting individuals against carcinogens.

Cancer experts estimate that 80–90 per cent of all human cancers might be linked to exposure to food additives, drugs, radiation, industrial chemicals and smoking. Many food additives that can cause cancer have been taken off the market and cigarettes have health warnings printed on the packages, but a large percentage of industrial chemicals have never been tested to determine if they are cancer-causing. Others, like formaldehyde which has caused cancer in laboratory animals and has been linked to illnesses in humans, are still being used widely.

Cancer is known to interfere with conception itself when it affects male and female reproductive organs. Men who have had frequent skin contact with cutting and lubricating oils on their jobs have developed cancers of the scrotum and male rubber workers seem to have higher than expected rates of cancer of the prostate. In addition, though very little research has been conducted in this area, a few studies seem to indicate a link between childhood cancer and the father's exposure to hydrocarbons and lead prior to conception.

Female cancer can involve the cervix, uterus or ovaries. Physicians advise cancer patients of both sexes who are undergoing radiation and chemotherapy not to attempt to conceive children during this time as these treatments are extremely hazardous to egg and sperm cells and to the developing foetus. Healthcare workers handling chemotherapy or who work in X-ray and nuclear medicine departments should also exert caution.

According to the new pregnancy legislation implementing the European Pregnancy Directive, all employers must assess their workplaces and, if hazards are found, must inform employees and decide what measures to take. These can range from temporary adjustments to working conditions to a temporary leave with all contractual employment rights including those related to pay to be maintained. (For more information on this legislation see Chapter 1 and Appendix D.)

MEDIA MESSAGES

Most of us get much of our information from newspapers, magazines, radio and television. The media often inform us about important hazards, but sometimes make us more anxious than we need to be. For our peace of mind, it is also helpful to understand how the mass media interpret and present risks to the public. We often assume these reports to be more accurate than they actually are, and we should remember that scary headlines sell newspapers and sources of advertising revenue colour interpretations. Often one article cites a substance as causing a specific illness and a later article denies it.

> During this pregnancy I was worried about an article in the newspaper about the water supply in this area causing a high risk of cancer because the pipes have asbestos linings. The first thing I thought of was 'What kind of water am I drinking?' A week later another article said that people shouldn't get upset about the water as the scientists weren't sure whether the water or something else was causing the increased risk.
>
> Clarissa, US operating room nurse

As Clarissa's experience indicates, media reports should be viewed as 'alerts' to search for further information rather than as 'facts' to be acted upon. But we should also be aware of certain specific biases that are an integral part of media reporting. Once we determine what these are, we can discount some of what we read and concentrate on deciding what further questions need to be answered for us to make an informed decision. Newspapers, magazines and television need to attract large audiences in order to survive financially. Increasing circulation and ratings are the name of the game. The media over-report dramatic events involving a large number of injuries and fatalities that occur at one time because these stories attract readers and viewers. The more common everyday risks and hazards causing an equivalent or even larger number of diseases, injuries, or premature deaths over a longer period of time do not get equal coverage. These occupational and environmental exposures may be of more concern to you if you are planning to become pregnant, but they are not news.

Journalists do not always have an easy time covering technical stories. They often do not have sufficient expertise in the specific areas they have to cover and they have to write under pressure to meet strict deadlines. It is often easier and faster, therefore, for them to obtain information from government offices or corporations that have public relations departments than to track down the individuals exposed to toxic substances and obtain their side of the story.

The media also indirectly take editorial positions by featuring stories dealing with those social issues and perspectives they deem important. Both sides of an issue are usually presented, but one is often highlighted and given more coverage.

You can find valuable information that contradicts the highly publicised version of an issue –you just have to look a little harder. Whereas mainstream newspapers and magazines appear on every news-stand, this other information can mainly be found in newsletters and magazines published by citizen, labour and women's groups of which you often have to be a member or pay for a subscription in order to get (newsletters published by the London Hazards Centre and City Centre are examples).

Women reading about reproductive hazards need to learn to detect biases when they are sifting and evaluating what they read. They must ask how complete and accurate is the reporting of risks and what sources were used. Most of the articles do not present sufficient information upon which to base an opinion. For those of you who want to investigate further, here are some questions to think about when you are reading an article on workplace exposure:

- Who was the source of the information? (e.g. the government, a corporation, a union, academic scientists, a women's group)
- Was the information checked with other sources?
- Was more than one perspective presented?
- How many studies was the article based on?
- Were the studies conducted on humans or animals?
- Does the article tell you who funded the study?
- Does the article tell you about the weaknesses of the study and whether more research is needed before the results should be acted upon?
- Does the article tell you where you can obtain further information about the scientific report?

You will find that most of this information is missing from stories in the media, but you can use the leads the articles provide to gather the additional facts you need to come to a decision about the workplace reproductive risks you face. This requires a good deal of dedication and time on your part because the kind of information you want can be difficult to trace, but with a little perseverance, it can be done and it is worth doing.

If you become concerned about the health of your workplace after the media has alerted you to a specific hazard, ask your GP, midwife or consultant to obtain information for you. Your primary healthcare provider can call the toxicology hotline which, because of the shortage of staff and money, can only take calls from healthcare professionals. As a concerned individual, however, you can call the London Hazards Centre or union health and safety offices (see the list of useful addresses at the end of the book).

Partially in response to the public's concerns about reproductive hazards (often generated by media coverage), birth defects data bases have been developed in many industrial countries and some have been linked into computer networks. Medical researchers share information and report their findings at international forums. It is hoped that these world-wide efforts will hasten an understanding of the causes of these conditions.

WHAT MAKES SOMETHING A HAZARD AND HOW IT CAN BE MEASURED

Before we can talk about hazards and risks we must define the terms. Frequently, the term 'hazard' refers to the substance or condition that is suspected of causing harm and the term 'risk' suggests the probability of harm arising from the individual's exposure to that substance or condition. For example, lead is a workplace hazard for a pregnant woman. The risk is the likelihood, or probability, of her having a miscarriage or giving birth to a child with a birth defect, as a result of her particular exposure. These are the definitions used in this book. Elsewhere, however, the two terms are used interchangeably. The term 'risk', for example, is sometimes loosely used to refer to harmful outcomes such as cancer, miscarriages or sterility.

There is a process by which scientists assess the level of risk. They:

- identify the hazard
- then identify the relationship between exposure to the hazard and its unhealthy effect on humans
- then try to determine the number of people suffering from this unhealthy effect and whether certain groups may be at greater risk.

For example, if exposure to X-rays is the issue, researchers would be particularly interested in investigating the effects on patients receiving high doses of X-rays and workers in occupations working with X-rays. Exposure is an important aspect to consider because many chemicals only exert harmful effects at high exposures.

Once a hazard is identified and scientists have studied its impact, those designing health and safety regulations have to decide whether to ban that substance entirely or determine an acceptable risk level. With regard to cancer-causing agents, a dose that gives a probability of having one animal in 100,000 develop a tumour over a lifetime is usually chosen as an acceptable risk level. For example, in the United States this is roughly the equivalent of 30 tumours per year assuming that the entire population was exposed to the same dosage as the animal and assuming that humans have the same sensitivity to the chemical as the test animals. This is then called a 'virtually safe dose' (VSD). In real life, of course, some people are more exposed than others and some are more vulnerable than others. Furthermore, these

VSDs are based on animal data and, as we saw in Chapter 2, animals are different from humans. Finally, these VSDs are estimates for cancer, not for reproductive harm, so a VSD can be used only as a crude guideline in estimating a risk if you're worried about the effects of the work environment on a pregnancy.

Part of the scientific argument about health effects also revolves around the question of whether there is a level of exposure below which there is no harm and, if so, what that level is. For example, very low doses of dangerous pesticides used to be considered safe. Now this is controversial. Due to limitations of both animal data and mathematical models, experts do not universally agree upon the answers.

Unfortunately, the estimates of a VSD depend not only on the data gained from the animal studies but also on the mathematical model chosen for translating the animal data into information that can be applied to humans. This means that if scientists use different mathematical models to interpret the same data, they will arrive at *different* VSDs which may be far apart. Faced with these diverse estimations, researchers concerned about the long term, as well as immediate, health effects on workers support the use of the mathematical model that estimates the VSD to be very low even though a higher dose might not be harmful. They prefer to err on the side of safety, believing that the best public health decision is to offer the greatest possible protection to the people exposed. This position is especially important as current scientific thought suggests that carcinogens do not have a threshold (an upper limit below which no cancers are produced).

But standards have to be set, and in the UK the COSHH regulations cover hazardous substances and agents in the workplace. In addition some of the regulations implementing EU work-related directives have set what are thought to be safe levels of exposure.

Even Scientists Can Only Make Informed Judgements

In many instances there is no firm proof that a certain amount of exposure to a specific toxic substance results in a specific degree of harm to our reproductive health or the health of our unborn children. Thus, even scientists have to make decisions based on *rough* estimates derived from human, animal, and bacterial studies. More is known about cancer-causing substances (carcinogens) than about substances affecting reproduction, and more information is available from animal studies than human studies. Recently some of the basic assumptions made in interpreting animal data in terms of the effect on humans has been questioned. In the spring of 1993 *The New York Times* reported that a good deal of evidence had accumulated indicating that chemicals often have different effects on animals than humans. Because of this officials sometimes do not act on the findings of studies based on rats and mice.

Occasionally substances are found to cause cancer because humans have been exposed to very high doses of them or lower doses for very long periods. This has happened with asbestos workers, uranium miners and soldiers participating in atomic testing.

Due to the uncertainty involved, it is especially important to understand what is meant by scientific and statistical evidence. Often this material is specialized and complex and research scientists seldom possess the expository skills needed to present the issues simply and clearly. Too often you need to read the explanation several times before it makes sense.

Managers also often have difficulty evaluating risks and hazards. When concerned about reproductive hazards, they focus mainly on pregnant women, forgetting about the male's role in procreation. Therefore, this knowledge will not only help you to evaluate general health hazards to yourself and your foetus but also to defend yourself against job discrimination masquerading as protection of your pregnancy.

We are only beginning to link occupational exposure of both parents to sterility, miscarriages, birth defects and childhood cancer, but even when the focus is only on the mother, much valuable information is not collected.

VARIOUS VIEWS OF RISKS AND HAZARDS

You will probably have different views about risks and hazards than scientists, business people or the Health and Safety Executive. You see risks in terms of your life and the lives of your future children who may be damaged by toxic substances in the workplace. You do not want to be the guinea pigs if insufficient testing is done before a chemical is allowed on the market. Even more important, you are concerned about the thousands of chemicals already available that either have not been tested at all or have been inadequately tested.

Risk-taking choices are based on your values as well as a supposedly rational calculation of costs and benefits. They are shaped by your own and other people's experiences with risk-taking and those you hear about on television or read about in magazines and newspapers.

Scientists, on the other hand, strive to maintain the integrity of their professional disciplines and require a high standard of proof. They want to be 95 or 99 per cent sure that their conclusion is not due to chance alone before they will state that a given substance does or does not cause harm (technically known as statistical significance.) Such information may take many years to obtain as this high level of proof depends on:

- the number of individuals in the study
- the strength of the toxic substance
- the amount of exposure to the substance
- the number of other factors that might cause the harmful effect
- the length of time the people are studied.

You, on the other hand, may want to protect yourself from possible reproductive harm when the available evidence is much weaker.

Business interests also often demand to 'go slow' because correction of hazards costs money. Business people often argue that the cost of the required safety features will force them to close their plants, and deny that there are involuntary risks. They point out that many people are willing to undertake risks if they are paid enough. Managers will argue that if they explain the risks, people then can decide themselves whether to take or avoid them. But there is a fallacy in this argument. Those workers who bear the risks may have little choice – job loss or a reduction in take-home pay may be too high a price to pay for a reduction in health hazards.

Some of you might be willing to accept hazard pay to improve the financial security of your families under certain circumstances, but this trade-off only makes the choice between work and health even more difficult as it highlights the risks you are taking. If you are pregnant or planning to become pregnant, this is not a good option.

The HSE and Employment Department are caught in the middle, pressured by the business and scientific communities who, for different reasons, advocate a cautious approach and by workers and consumers who demand that they act immediately. Each side uses research studies to buttress its own position. Perhaps what is needed is a new interim standard of evaluating substances or agents suspected of causing occupational health hazards to protect workers. What makes sense for scientific purposes may not make sense when you are trying to protect your well-being and that of your future child.

WHAT RISKS SHOULD YOU TAKE?

In our society pregnant women bear the primary responsibility for the health of their foetuses. Thus, concern about the outcome of their pregnancies takes a high toll in anxiety as well as illness. In order to reduce their fears, they can learn how to rank hazards they face in terms of severity and manageability so they can decide which ones they are going to try to do something about. For some hazards, simple precautions can be taken such as washing your hands after coming into contact with a substance that you suspect of being toxic, opening a window or checking the ventilation system to make sure that it is working properly if you think you are suffering from indoor air pollution.

Without some basis for judgement, however, hazards you choose to avoid may not be those that do the most harm. You need to keep things in perspective: *remember that the overwhelming number of women deliver normal babies.* Constant worrying may turn out to be as harmful to your pregnancy as a brief exposure to a low level of a substance that is not highly toxic.

Evaluating the risks you face may seem like an impossible task because of the amount of uncertainty involved, but that is not so. Research on the accuracy of public perceptions of health risks shows that consumers have a surprisingly good idea of the relative frequency of most causes of death although their knowledge about specific risks may not be precise. Workers considering becoming pregnant are even more concerned about health risks than the general public and are likely to improve their risk estimating capabilities if presented with clear and accurate information.

WHAT ARE YOUR OPTIONS?

What you decide to do with information you gather depends on your individual situation. Experts cannot tell you what the right choice is when your only options involve some degree of hardship or risk. They can only provide information. You are the one who has to live with your decisions. Nobody can provide all the answers, but by helping your healthcare provider in identifying and evaluating possible harm due to workplace hazards, your GP or midwife should be in a better position to ensure you have a healthy pregnancy.

Finding Out Whether You Are at Risk From an Occupational Hazard

You should try to find out the following.

- Which substances you are exposed to in the workplace.
- What kind of health problems can be caused by these substances? Some individuals have allergies to substances that do not seem to have ill effects on others. If you suffer from any allergies and are pregnant, be sure to question carefully about possible kinds of allergic reactions.
- Do you breathe, swallow or have skin contact with these substances?
- For how long have you been exposed to these substances – minutes, hours, days, months, years? Is the exposure continuous or occasional?

By knowing about the risks attached to reproductive hazards, you will be able to make some kind of choice even if it is only whether you stay in the job or leave. If you are thinking about getting pregnant, be a careful listener. You can often learn a great deal about what is going on around the workplace by listening to co-workers' complaints. Hazards in the workplace may be hidden. It may only be when someone else suffers the ill effects that you learn that you too are at risk.

> I know of one incident where a woman was working with a lot of chemicals in the electronics company. She had quite a few miscarriages. Her doctor had to tell her it was related to all the chemicals she was working with. Two other pregnant women were not aware of any hazard until their doctor told them and then they did go out on disability. They had trouble collecting from the

> company which wasn't so willing to admit that it was more or less their negligence.
>
> Caroline, US electronic microcircuit wirer

Pregnant women working under conditions similar to Caroline's may decide to leave work rather than face the risk of harming their foetus. Under the maternity rights legislation that came into force in 1994 you should not have to make this choice. Where your job cannot be made safe, your employer must give you suitable alternative work or suspend you on full pay for the period needed to protect you and your foetus.

> I wore a badge and every month we took it off and sent it out to the lab to make sure I wasn't getting any radiation. The badge was always negative. If there was any trace of radiation, I would have had to terminate my employment.
>
> Rory, US dental assistant

Rather than leaving or invoking their legislative rights, many pregnant women prefer to work out informal arrangements that allow them to avoid those aspects of their jobs they feel are dangerous during their pregnancies. If your supervisor is sympathetic and co-workers willing, this solution often works well.

> My job required travelling about 50 per cent of the time. I wanted to continue working but was afraid that so much travelling would harm the foetus, especially as I had suffered prior miscarriages. There was no way that I could keep the job and not travel for my entire pregnancy. I spoke to my doctor and we worked out a compromise. I would not travel past my seventh month and I would allow extra time for travelling the rest of the time to reduce stress. The doctor also suggested that I always carry around one pound coins and single dollar bills for tipping porters so that I never had to lift anything heavy. This worked out well.
>
> Laura, marketing director

> If my legs were tired or swollen, I would sit with my feet up. People would make allowance because of my condition. When I was teaching a workshop, I would sit down unless somebody wanted

> help at the machine. The students would come to me if it didn't require anything to do with the machine. Ordinarily, I would walk around all the time. If my feet were swollen when I had to lecture, I would lecture sitting down with my feet on the table. I told them why I had to do this. Otherwise they would think I was lazy, and I didn't want them to think it was normally all right to work like this.
>
> Kate, college lecturer

Still others ask for transfers to other kinds of jobs for the duration of their pregnancies, or for a leave. Even when options are open, you often do not know which ones to take. Although the law gives you certain rights, some employers may make life difficult for you if you try to use your rights, even trying to fire you on grounds other than your pregnancy and being willing to lie in the written statement of explanation required under the current legislation. Sometimes it is only when you are trying to return after maternity leave that you realise that your employer may only obey the letter, but certainly not the spirit, of the law.

Making sure that the law protects you and is strongly enforced is your best security.

THE EXPERTS DON'T AGREE

The roles scientists, consumers, businesses and the government play in the risk evaluation process is very complicated. Legislation usually ends up as a political compromise between parties who have different interests – in this case employers, employees and governmental regulators. The area of occupational health, however, is even more complicated as there is also disagreement among the experts. For example, even when there is agreement on risk assessment, there may be no agreement on what is an acceptable level of risk for individuals to face. How many individuals have to suffer ill effects before government action is taken to ban or limit the substance? Unfortunately, some of the answers are based more on the amount of money it costs to clean up the workplace rather than on the extent of the risks involved. Numerous suggestions for determining acceptable risk levels have been proposed. Some are based on a standard of relative risk rather than absolute risk and recommend that risks from comparable activities should be used as a bench mark for acceptable risk. For example, Lord Rothschild proposed that a risk of exposure to a cancer-producing substance should be considered unacceptable when the chances of death are above one per 7500 a year, the risk of being killed in a car accident in Great Britain. A better idea may be to make both automobile driving and the workplace safer.

Some of you might want to delve more deeply into the workplace's influence on your reproductive health. According to the current UK health and safety regulations, all employers must assess their workplaces and if hazards are found, they must inform their employees and decide what measures to take. If you are not reassured by the assessment and still feel unsure about the risk of working in a job that you suspect might have some reproductive hazards, you can also discuss your concern with your union safety representative, members of the occupational safety and health staff at work, or call or write to the Health and Safety Executive, or the Maternity Alliance. The addresses and phone numbers of helpful organisations, unions and government offices are listed at the back of the book. The appendix to this chapter also explains in detail the various kinds of animal, bacterial and human studies researchers conduct to determine whether a substance or agent is a reproductive toxic.

CHAPTER 4 APPENDIX 1: CANCER AND REPRODUCTIVE STUDIES

Cancer Studies

In cancer studies, the mice or rats in the experimental group are exposed to specific chemical doses during their lifetimes of 24–36 months. Generally the lowest dose used bears some relationship to anticipated human exposures while the highest dose may be hundreds or thousands of times higher. Another group of mice or rats act as controls; they are treated exactly the same as the animals in the experimental group but not exposed at all to the chemical being studied. Scientists usually call this a zero dose. At the end of a cancer study, the scientists count the number of tumours or lesions that grew in the animals receiving the different doses of the chemical substance and compare these results with the number of tumours found in the control group. For example, if 10 per cent of the rats in the control group but 50 per cent of the rats in the experimental groups develop tumours, the large difference would make the researchers fairly certain that there were toxic effects. If, however, 10 per cent of the rats in the control group and 13 per cent of the rats in the experimental groups develop tumours, it would be harder to pin the blame on the toxic substance being investigated.

Cancer studies take several years and are very expensive. The care and feeding of the experimental and control groups of rodents is a tremendous task and may rival childrearing in sleepless nights and unexpected hazards. Sometimes the investigation requires that 700 rats be fed and tended and as many as 500 of them receive measured doses of substances every day for two years. Laboratory workers must record the general condition of the rats every day and weigh them twice a week. Every rat that dies before the experiment is concluded must be examined, its death recorded, and samples of its tissue stored for future analysis.

At the end of the study, each surviving animal is killed; the main organs are analysed and stored in formalin, and tissue must be prepared for microscopic examination. A typical two-year study can generate nearly a million pieces of information plus 250,000 slides.

With this amount of work to complete, lab technicians can make mistakes in many different ways. Before international agreements on good laboratory practice standards came into effect, occasional horror stories emerged. Water bottles remained empty for days because the workers didn't come in at weekends and nights when they were supposed to, or rats got sick and died because lab workers were too busy to clean the cages properly, allowing bacteria to multiply.

So, how accurate are the records produced by these studies? The recently instituted Good Laboratory Practice Standards include inspections, so the previous laxity has been for the most part eliminated. Many chemical substances still on the market, however, are deemed to be safe on the basis of studies conducted before the standards went into effect – if they have been tested at all.

Drawing conclusions about humans from test results on animals is especially difficult because some of the animals routinely develop tumours that are not caused by the substance being tested.

Reproductive Studies

Reproductive studies are usually conducted over a much shorter time span than cancer studies. In a reproductive study, the investigators concentrate on reproductive markers such as birth defects or still births in the litters or pregnancy rates of the females in the experimental groups compared with those in the control group. The short pregnancy of 22–23 days is one of the reasons that mice and rats are particularly suited to be experimental animals for reproductive studies.

These studies are generally of two types. First there is a general reproductive and fertility test which investigates the effect of a suspected toxic substance on fertility and pregnancy and its transfer through the mother's milk. The second kind is called a teratology test and is specifically designed to determine the effect on foetal development.

General Reproductive–Fertility Studies

In general reproductive–fertility studies the mice or rats are usually divided into three experimental groups, each receiving a different dose of the substance, and one control group. The substance can be injected, rubbed on the skin or combined with food. The control group receives either nothing or a placebo (an inert substance).

The treatment of males starts 60–70 days before mating. This is their sperm cycle – the length of time that sperm take to mature fully from stem cells to 'adult' sperm cells capable of uniting with the egg and forming an embryo.

The female mice or rats are treated with the decided upon doses of toxic material 14 days before mating. This covers three to four ovulation cycles. At the end of the 14 days, the treated females are mated with the treated males. Pregnancy begins on day 1 when the egg is fertilised; implantation (when the fertilised egg embeds in the womb) occurs on day 6. The period when the organs and structure develop (embryogenesis) occurs from day 6 to day 15. Birth occurs at 22–23 days. The offspring are 'nursed' by the mother during the three-week period of lactation, at the end of which they are weaned.

As we know, infertility is a fact of life for too many young couples. Controlled experiments ask whether chemical substances can be blamed by comparing the number of successful matings within each treatment group and the number with the control group. If scientists want to know whether the fertility of the male or the female is affected, they can mate one male with two females, one from the experimental group and one from the control group. If neither female becomes pregnant, the problem probably lies with the male, but if only one female becomes pregnant then the toxic substance is likely to have affected the female. If both become pregnant, fertility has not been influenced by the treatment.

Miscarriages and breast milk pollution are two additional worries of pregnant workers. Again, controlled animal experiments can help shed light on these aspects of reproductive health. If the female delivers a normal-sized litter and the offspring stay alive until weaning, this gives a measure of the mother's ability to carry to term and to feed her young successfully. If the baby mice and rats start dying off during the nursing period, then toxic effects are most likely being transmitted through the mother's milk. The remaining offspring can also be killed at weaning and autopsied in order to ascertain whether any further damage has been done.

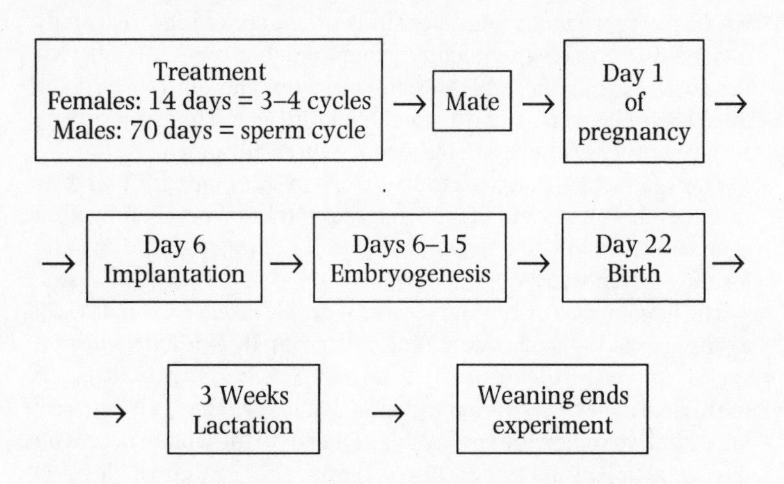

It would seem that malformations could also be detected by these reproductive–fertility experiments, but this is not the case. The mother rodent usually eats a defective offspring immediately after its birth, and births often occur at night when lab workers are not in attendance. Therefore another type of experiment called the teratology test has been designed. In this kind of experiment, neither the male nor the female in any of the treatment groups is exposed to the toxic substance before mating or during the period of pregnancy before implantation.

Teratology Test

The different doses of the toxic treatments are started for each group during the developmental period (embryogenesis) from day 6 until day 15. On day 21 (the day before expected birth), the pregnant rat or mouse is killed and the foetuses analysed. While this is hard on mother rat, much worthwhile data can be gathered. Because the dead mother cannot eat the malformed foetuses, the researcher can count how many there are and what kind of defects they have. The researcher can also look for little blobs of tissue that indicate a reabsorption – equivalent to a spontaneous abortion in humans. If the embryo or young foetus dies in the uterus, it is reabsorbed rather than expelled. Stillborns can also be counted. These are foetuses that died too close to the due date to be reabsorbed. Finally, each live foetus can be weighed and their weights compared with the dose level of the toxic substance received by the mother. These four measures – malformations, reabsorptions, stillbirths and body weight – are compared between all of the three experimental groups and the control group.

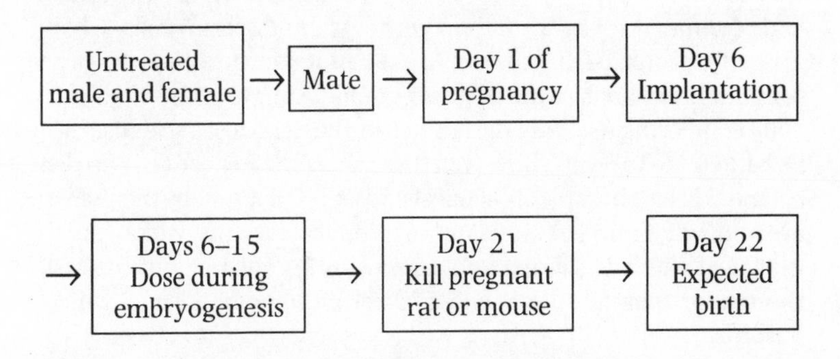

We can learn a lot from such experiments but, as is true with cancer studies, whether the findings apply to humans is not always clear. In general, scientists feel more certain that these results would apply to humans if the same results were obtained with two or more kinds of animals, e.g. rabbits and mice.

Bacterial Studies

Scientists have been slow to study male and female workplace-related reproductive hazards, but new tests using living bacterial or animal cells instead of live animals have stimulated investigations in this area. It is hard to imagine that a lab technician can examine a billion bacteria in a three-square-inch area, but that is what is happening. Many scientists are excited about these types of studies because such large numbers of living organisms can be examined so quickly. Genetic material is similar in all living organisms and millions of bacteria can be studied at one time. These large numbers are needed in order to find evidence of mutations which are fairly rare and which could trigger cancers that would normally take many years to develop in whole animals.

Although the development of cellular tests has excited many researchers, others are more cautious. They believe that while these tests are quicker and cheaper, they are not necessarily better predictors of human toxicity. Bacterial studies can only test for genotoxics (substances affecting genetic material in the cells), but these are key and many scientists believe that if they could identify, eliminate or at least reduce the use of the important genotoxics, they could go a long way towards cleaning up the workplace. Genotoxics usually attack body (somatic) cells rather than germ (ova and sperm) cells, which are better protected. If, however, these genotoxics do attack the germ cells malformations in the offspring could result.

By comparing results of the animal and bacteria tests, scientists will be better able to decide how well these shorter, cheaper tests predict potential health hazards of chemicals. So far cellular tests by themselves are somewhat unreliable and may overpredict hazards, which means that even with all its drawbacks animal testing still remains the best, though far from adequate, way of determining long-term chemical hazards.

Human Studies

Human research in occupational health involves what are called epidemiological studies. These are primarily of two kinds – *retrospective* and *prospective*. Take the case of researchers who are investigating the relationship between miscarriages and exposure to anaesthetic gases that leak into the operating room.

In a retrospective study, the researchers would compare nurses that had miscarriages with nurses that didn't, making allowances for other possible factors that might account for the outcome such as age, number of previous children, history of miscarriages in the family, number of years employed, possible exposure to other hazardous material and so on. They would see whether a higher percentage of nurses who miscarried worked in the operating room than worked in other areas of the hospital.

In a prospective study, researchers would compare the reproductive patterns of a matched group of operating room and non-operating room nurses over a number of years to determine whether the miscarriage rate was higher among the operating room nurses than among nurses working elsewhere in the hospital.

5

Seeking Solutions

We have come a long way since the factory girls laboured under unsafe and unsanitary conditions for four cents an hour. As we have seen, however, subtle types of harassment and discrimination have accompanied these improvements. These hidden agendas are more difficult to prove and in some ways more difficult to combat as the practices are defended on other grounds – efficiency, protection, equity – all of which sound laudable. Pregnancy or just the biological potential for pregnancy has been used in the past, and continues to be used, by some employers as the unstated excuse for policies that impede equality, job security and career mobility for women.

But despite some of these obstacles, you *can* make your workplace better. You can improve your working conditions in many different ways to protect your own health and that of your future children. Some involve major changes in policy and expenditure while others only require small substitutions that employers might be quite willing to make if these were brought to their attention. Some initiatives you can take yourselves, others you can ask your employers to undertake, and some changes need to be initiated by your union or a government department.

Things to do

What you can initiate yourself

- Educate yourself about the hazards you face.
- Form your own worker and/or community action group.
- Get involved in 'barefoot epidemiology' – collect your own evidence.
- Organise around the workplace hazard involved.

- Fight for wider interpretation and greater enforcement of the health and safety regulations.

What you can ask your employer to do

- Provide a health and safety course so that you know what hazards to look for.
- Provide nutritious food in the cafeteria or a nutritious snack machine.
- Provide a place to relax during breaks.
- Provide flexitime schedules, parental leave and subsidised childcare arrangements.

What you can ask the Health and Safety Executive or union to do

- Press for further research on substances suspected of being toxic and their non-toxic substitutes.

All of you have individual needs. Your workplaces pose different kinds of problems and your co-workers and employers provide differing degrees of support. If you are lucky, colleagues – males as well as females – can be a major asset in improving your work life when you are pregnant. If you are unlucky, as are some of the pregnant workers described in this chapter, you may be harassed or discriminated against. Each one of you also has different amounts of time and energy that you can spend on actively improving your workplace conditions. Many of you may feel uncomfortable about being assertive about your rights – a reflection of the way in which many women have been brought up. You may be brave in theory but become anxious when it comes to confronting supervisors and employers. You do not have to be Superwoman. Engage in the activities that are comfortable for you.

This chapter provides some easy solutions that will make your work day healthier and less stressful. It also presents ways to become more involved and active in the struggle for a reproductively safe work environment. If possible, include men in your organising and action efforts. If this is not feasible then women on their own can be a potent force. Most of the women who have become involved in this sort of struggle started out with no special background or training, they simply became angry at the conditions they worked in and the way they were treated. They were amazed at how quickly they could learn the ropes and how effective they could be.

If you do not have a safety representative in your workplace who can do it on your behalf, one thing that many of you may want to do is to carry out your own general health and reproductive health evaluation of your workplace exposure to toxins. The following list of questions may be helpful in doing so:

- If a substance is harmful to your health
  - can it damage the foetus?
  - can it cause a mutation?
  - can it damage specific organs of the body?
  - can it cause cancer?
- If it is harmful, how harmful is it?
- If it is harmful, what is the likelihood that you are exposed to it?
- Who may be particularly at risk?
  - pregnant women
  - foetuses
  - individuals working in certain occupations
  - individuals living in certain communities
  - children
  - men and women contemplating having a child
- What kind of evidence was the risk assessment based on and how strong is it?
- What were the sources of information on which the estimates were based?
- How much uncertainty is involved in the estimations?
- Upon what assumptions are the estimates based?

Remember, however, that before you approach your management seeking improvements, it is a good idea to do some extra homework and think of ways in which you can present your suggestions in terms of long-term savings, for example increased efficiency and decreased injury and absenteeism for all workers, including pregnant ones. Ease of maintenance and inexpensiveness of repairs is a prime consideration for the employer who is considering making improvements in the workplace that would make you more comfortable.

Furthermore, pregnancy can become a catalyst for action. Women will often organise around reproductive threats in the workplace and community more readily than they will for other serious health hazards. For example, if inadequate sanitary and safety precautions are typical of your workplace, keep records of the specific incidences documenting any accidents, injuries and near misses and give them to your safety representatives. Even if you are not unionised or have no safety representative, familiarise yourself with the laws designed to protect you. Brief descriptions of the Management of Health and Safety (MHSW) Regulations 1992 and other health and safety laws are in Appendix C at the back of this book. Appendix E contains summaries of the European Union Work Related Directives.

SUCCESS STORIES

While it is easier to fight for your rights with a union or safety representative backing your claims, throughout the world, women by themselves, in groups and in coalitions with other interested parties have exerted pressure on employers and governments to reduce health hazards in their workplaces and in their communities. In a few cases, women even received national coverage for their efforts. They brought to the attention of the public and the government severe problems such as the leaching of chemicals in Love Canal in the United States and, in the UK, radiation leaks attributed to the Sellafield nuclear reprocessing plant and repetitive strain injury (RSI) caused by work practices at British Telecom (BT).

Yet even smaller, less heralded efforts have made a difference. More women would organise for change if these stories were told and 'know-how' transmitted. Women have been very ingenious and creative in the ways they have gone about improving the comfort and safety of their workplaces. Furthermore, they have developed techniques and approaches to be used by both the shy and retiring and the more assertive female workers.

Women throughout the world have used legal suits (both individual and mass action), lobbying and legislative reforms, community action and community health programmes, worker solidarity actions, and research to promote change. In Italy, for example, women have drawn up a new legislative proposal to restructure the work day and work week, vacation time and other components of employment to improve the lives of working women and their families.

The Latin American and Caribbean Women's Health Network reports on some of the various initiatives taken in their countries. In a Mexican lawsuit, women who were pregnant while working at an electronics plant are suing the company for negligence in not warning them about potential dangers posed by chemicals they were working with and for failing to provide the necessary protective equipment. These women suffered miscarriages, stillbirths, and had babies with severe physical and mental disabilities or who died shortly after birth, which they attributed to their workplace exposure. Even male-linked reproductive hazards are receiving attention. In Costa Rica, some 150 men who were employed on a banana plantation sued the Standard Fruit company, asserting that their sterility was caused by use of the pesticide DBCP, a known reproductive hazard which has been banned in the United States.

In the UK, the Equal Opportunities Commission has been arguing for better working conditions, better pregnancy, maternity leave and family leave policies and more and better childcare options and sites. Women have lobbied within unions to highlight women's workplace problems and to have them addressed. The Trades Union Congress (TUC) publishes some useful booklets for working women (contact them at Congress House, Great Russell Street, London WC1B 3LS) and in its equal opportunity campaigns not only focuses on hiring more women, minority workers and workers with disabilities, but also includes within the equal opportunities concept negotiations toward getting good agreements in areas such as:

- childcare, including after-school and school holiday schemes
- maternity and paternity rights
- training
- return to work rights following childrearing
- sexual harassment
- pensions
- family leave.

Within many unions, women's health groups have been formed. They often try to determine what chemical substance, physical or biological agent they are exposed to and what its effects are. In a tennis ball factory in Sheffield, assembly line workers, mostly women, used a chemical solvent to treat the cloth used in the manufacture of the balls. The women complained of dizziness, weakness, headaches and other illnesses. An industrial chemist was brought in as a consultant and attributed the problem to 'mass hysteria', a term seldom used to describe health complaints made by men. It was only after 50 women collapsed and ambulances had to be called that this section of the plant was closed down.

The women's health group went to work and discovered that exposure to the solvent was linked to cancer of the cervix, brain cell damage, infertility among women and men and personality changes. The company eventually agreed to pay £4000 to those most severely affected and £100 to those suffering more minor discomforts.

Eleven women who suffered from RSI due to their keyboarding jobs at British Telecommunications (BT) also were pathbreakers. They, with the support of the National Communications Union, sued BT for compensation for their injuries. The women were ordered to type 3.6 letters per second, and if they did not reach the target, the computers automatically docked their pay and informed management. To make matters worse, the women were given chairs that were not adjustable and at times were using drawers as desks and bathroom scales as footrests. Despite these horrendous conditions, it took a six-year battle to reach the courts. The court ruled in favour of the women, but only deemed BT negligent with respect to the workstations and not for their long hours and monitoring practices. BT was planning to appeal the decision but agreed to an out-of-court settlement for an undisclosed amount of money. What is most important, is that the original court judgement finding BT liable because they failed to provide suitable workstations still stands. The new regulations protecting VDU workers now provide for rest breaks.

The government also uses the plight of women workers in their public relations efforts. The former prime minister John Major's publicity campaign in launching Opportunity 2000, the business-led initiative aimed to improve the position of women in the workplace was, at least indirectly, in response to the continuing pressure women exerted for better and more flexible work opportunities. Top executives seem to have got the message – even the Bank of England is conducting an audit of its jobs to see where schedule flexibility is an option. Unfortunately, it is frequently the male line managers who make the lower level decisions that remain prejudiced against (or threatened by) women workers. Women who answered a survey conducted by the Institute of Management felt that 'the old boy network' proved to be a greater obstacle to their advancement at work than did inadequate childcare or insufficient opportunity for a flexible work schedule.

WOMEN'S INITIATIVES IN THE UNITED STATES AND CANADA

In the United States several different kinds of initiatives have been undertaken. In one company, women were having breathing problems which they suspected were caused by polluted air in their offices. They contacted the staff of the occupational safety and health programme of their union, the Communication Workers of America (CWA). The CWA representative and the women involved investigated the problem together. First, they requested information from the employer's files about the chemicals used in the ventilation system and discovered that freon, a highly toxic solvent, was being used. They conducted a simple survey which documented the substantial number of workers affected and pinpointed the location of the workstations and air vents.

Next, the CWA brought in a health expert to conduct tests. Backed by the results of the tests and their own evidence, the women conducted informational picketing in front of the building. This led to bad publicity for the company and the management soon relented and installed a new, more efficient ventilating system. Most important, the new system used a less toxic solvent.

In another office the CWA, by filing grievances with the Occupational Safety and Health Administration (OSHA), convinced the employer to finance a health and safety education programme. The women wanted to learn how to identify hazards themselves so that these could be corrected immediately. By eliminating the time lag involved when the governmental agency was called in to investigate and confirm the hazard, the women would hasten the process. The company partly agreed to this request because it preferred not being the target of government inspectors. After taking the health and safety classes, one group of women was able to identify ozone emissions from a copying machine and tried to persuade the management to switch to another type of machine designed to prevent ozone formation. They were not successful, but did get management to improve the ventilating system, a second best solution.

One of the most effective demonstrations of women power in the office occurred at Boston University's Business School, where members of the United Automobile Workers refused to unpack the new video terminals until management provided ergonomically sound workstations similar to those installed in other departments. The unpacked boxes stood there for six months until the women's needs were met.

Sometimes pregnant workers are able to accomplish a good deal on their own. The National Action Committee on the Status of Women in Canada has collected several case histories, including Tess, a pregnant laboratory technician who worked in a non-unionised battery factory. Her job was to test samples of soil from the work site. When she discovered that they showed mercury contamination, she asked her doctor about the risk to the foetus posed by her exposure to mercury vapour. Her doctor warned her not to have any more exposure to mercury throughout the rest of her pregnancy because mercury vapour can be transported across the placenta inflicting possible damage to parts of the developing brain and nervous system of the foetus.

Tess requested a transfer to a mercury-free area of the plant. She was offered several positions that had less mercury exposure, but none at a zero level. She refused these. The company responded by asking her to sign an indemnity form which absolved the company from any responsibility for harm to herself or her unborn child if she continued working. Tess refused to sign. She finally negotiated a transfer for the remainder of her pregnancy to an office job which had zero exposure to mercury.

Tess's story had a happy ending, but sometimes, however, you ask the right questions and get the wrong answers. Take the case of Diana whose Canadian employer embarked on extensive renovations involving paint varnishes and glues in the area in which she worked. When she asked if these could be harmful to her pregnancy, she was assured that there was no danger. Three weeks after the work was started, Diana miscarried. She requested a copy of the data sheets listing chemical ingredients to show her doctor. Her physician felt that the presence of toluene in the paint and exposure to several other chemicals had caused her miscarriage.

Diana felt particularly betrayed because she had repeatedly asked for information about the fumes and was repeatedly assured that they were harmless. She then charged her employer with violation of the Canadian Occupational Health and Safety Act (the Canadian Act allows individual prosecutions). These legal remedies will not bring Diana's unborn child back, but they should make the employer more careful in the future and help protect employees' future pregnancies. Cases like Diana's show why protective legislation is so important. In the UK, the MHSW and COSHH regulations, if properly enforced, should protect a pregnant worker in Diana's situation. These regulations require that employees be informed about

- the risks arising from their work
- the precautions to be taken, and if carried out, the results of monitoring and the results of health surveillance.

But the law is only valuable if it is used and enforced.

When work situations are appalling and discrimination rampant, a courageous woman can find the strength to fight back under incredibly difficult circumstances and become a pathbreaker for the rest of us. Bonita is one such heroine. As a pump tender at the Steel Company of Canada, she suffered six years of unbelievably vicious harassment simply because she was a woman in a traditionally male job, and because she often filed complaints about safety and health hazards. For example, a gas booster that she complained was dangerous later exploded. Instead of being applauded for her actions, she was called stupid, accused of theft, her work competence was questioned and she was threatened with layoffs and the elimination of her job. She was used as a scapegoat to intimidate the remaining women at the plant. The company sank so low in its tactics that it refused to screen off the men's shower and changing room. Bonita was forced to see the men naked day after day. When she complained, she was told by her supervisor to shut up and enjoy the free show!

Bonita did not enjoy the free show. Instead she went public and filed charges against the company listing over 60 instances of abusive treatment, discrimination, sexual harassment, reprisals from management and unhealthy and unsafe working conditions. She demanded a far-ranging series of remedies and her complaint was precedent setting. It sought both to get the health and safety legislation to cover sexual and gender harassment and to compel companies to properly accommodate their facilities and procedures to women workers in non-traditional job areas.

SMALL WAYS TO START

Not many of us could persevere under the conditions Bonita faced. Luckily, they represent only one end of the spectrum – albeit an end that shouldn't exist. But most women can propose some small innovation to improve their working lives during pregnancy, and can then build on their increased knowledge, competence and self-confidence. It is sometimes effective to start in a small way in an area that is not likely to make waves and one in which their non-pregnant colleagues concur.

For example, after decades of unconcern, many workers are beginning to eat healthier foods and participate in physical fitness activities. Some large corporations are offering programmes to improve the health of their workers and reduce absenteeism – a trend from which all pregnant workers can benefit. The following are suggestions you might consider making to your employer.

- If your company is considering such a programme, ask those in charge to include an exercise class geared to pregnant women.
- If there is a company cafeteria ask that it provide fresh fruit, salads, juices, and milk if it does not do so already; that it opens for breakfast a half hour before the start of the workday so that you do not have to rush to eat at home or snack on cakes in your break.
- If there is no cafeteria, suggest that the employer looks into the competitive food service companies which provide meals and snack food on a contractual basis or, at the very least, provides vending machines that carry healthy sandwiches, fruit and yoghurt.
- If you work in a small office that cannot hire food services or open a cafeteria, suggest that at a minimum expense your employer could furnish a small area with comfortable chairs, good lighting, hot and cold running water and a small refrigerator and which could be used as an employees' lounge and lunch room. You can then put up your feet, reduce the pressure on your legs and relax. Socialising in pleasant surroundings reduces tension and fatigue and as a result will improve efficiency.
- If you are unionised you could ask your union to include flexible break schedules for pregnant workers in contract negotiations. If you are not, and your employer is not particularly sympathetic, suggest that you take a shorter lunch break or come to work fifteen minutes earlier in exchange for an extra break during the day. You could alternatively use holiday or sick leave for this purpose. Trading one eight-hour annual leave or sick day for thirty-two

days with an additional fifteen minute rest period might be a valuable exchange.

THE PREGNANT WORKER'S WISH LIST

If you are pregnant or planning to become pregnant and to continue working, set your desired goals. Then you can devise tactics toward reaching them. For example, you may aim to achieve the following from your employers:

- commitment to eliminating or reducing reproductive hazards in the workplace
- incorporation of work and family-life commitment into company policy and public statements
- implementation of hazard monitoring procedures and a grievance process
- strict enforcement of the legislation implementing the EU Pregnant Workers Directive and other work-related directives
- support for research efforts to assess reproductive hazards of the workplace
- provision of flexitime schedules
- provision of an on-site childcare centre
- provision of a childcare referral service
- provision of job sharing opportunities.

The Swedes have established the kinds of reproductive health standards and maternity benefits which we all want and hope we will eventually receive. But, unfortunately, the UK is still far from reaching such a goal. The new health and safety regulations if applied stringently should act as a catalyst, but unless these regulations are enforced and violators severely penalised, further progress will be in name only.

Increased protective legislation and more research, while badly needed, do not by themselves provide a solution to reproductive hazards in the workplace. Along with the striking gaps in knowledge about reproductive health hazards in the workplace are the gaps in communication and cooperation among those that should be working together at alleviating the dangers – workers, affected communities, trade unions, employers, clinicians, researchers and policy makers.

One way to make inroads in the area of communication and cooperation is to find interests in common with other workers so that pregnancy is not viewed as an aberrant condition – and there are fundamental ones. Men want their partners' pregnancies to be protected and non-pregnant women want their future pregnancies or those of their family members to be protected. Remember that if a workplace substance harms reproduction it also has other harmful effects on the health of all workers. Management personnel are not immune to these risks – they too are exposed in their workplaces.

Carrying out the tasks necessary to reach reproductive health goals in the workplace requires a great deal of time and energy. Pregnant working women and working mothers of small children have little extra time or energy. Therefore, informal alliances between women's health groups, unions and environmental groups are also useful in pressing companies and government agencies to work harder to implement a healthy reproductive workplace and a healthy work and family policy for all.

YOU CAN BE A 'BAREFOOT EPIDEMIOLOGIST'!

'What is a barefoot epidemiologist,' you are probably thinking, 'and why would I want to be one?' 'Barefoot epidemiology' is the name given to the gathering of evidence by women who believe that their health or that of family and friends have been harmed by exposure to toxins. Two of the most well-known cases of barefoot epidemiology took place in Love Canal, where women refused to accept New York State's conclusions that living in their homes did not pose any reproductive risks, and Woburn, Massachusetts where women traced an increase in children's cancer and leukaemia to toxic wastes from two major industrial companies. As mentioned earlier, women who lived in the area around Sellafield also were influential in getting sceptical officials to investigate the increase in number of childhood leukaemia cases. Most of these women were ordinary housewives, with no special training, who never dreamed that they would become so involved. They were initially propelled by frustration and anger.

A more recent example involved a woman pharmacist in an American small town in Louisiana who became concerned because a number of her sisters, friends and neighbours had become pregnant at the same time and all had miscarried. When she finished her inquiry, she found that one out of three pregnancies in her community did not lead to a live birth, more than twice the average for Louisiana. Her list was taken seriously because the town she lives in lies in the midst of an industrial corridor between Baton Rouge and New Orleans, which produces one fifth of America's petrochemicals and has one of the highest cancer rates in the country. The air, ground and water are so full of carcinogens, mutagens and embryotoxins that a union leader pithily referred to it as 'the national sacrifice zone'.

A few American, Canadian and British studies show an abnormally high rate of spontaneous abortions and birth defects among women whose husbands' work exposes them to vinyl chloride or who live downwind from vinyl chloride polymerisation plants – the situation in the little Louisiana town. Scientists at the Tulane School of Environmental Health are now searching for the reason behind this high incidence of miscarriages.

THE SELF-DISCOVERY OF WORKPLACE HAZARDS

Workers' self-reports may be the first indication that anything is wrong with the reproductive health of male and female workers. You may know more than you think you do. In fact one union leader in the United States reported that in all the cases he knew about, experts called in to investigate upheld the workers' allegations that there was a problem.

Probably the most dramatic case of worker-reported reproductive difficulty occurred in the late 1970s when several seemingly innocuous conversations led to the eventual banning of the chemical DBCP because it was toxic to the testicles. Workers in a California agricultural chemical plant and their wives noticed that no one in the group was having babies. It soon became evident that this was not by choice but a result of the inability of the couples to conceive. The workers became worried and informed their union, who asked university researchers to investigate the situation. The researchers confirmed the association between exposure to DBCP (dibromochloropropane) and decreased sperm count. The testicular toxicity of DBCP in animals had been known for more than 15 years, but without the couples themselves pushing for an investigation, it is unlikely that studies demonstrating the human connection would have been initiated.

Another group of workers exposed to halogenated hydrocarbons questioned what they thought to be an unusually large number of infant deaths. A preliminary epidemiological survey tentatively confirmed this connection. The UK does provide a few established mechanisms to improve workplace conditions, some of which are worth trying, depending on the circumstances.

ACTION THROUGH OFFICIAL BODIES

Action Through the Health and Safety Executive

A safety representative can call in the Heath and Safety Executive Inspectorate to enforce the law, but if you don't have a safety representative and you believe your employer is breaking the law you can contact the HSE yourself. Traditionally, the HSE and Local Authority enforcement officers have tended to see their role as advisers and educators, even when they are called to investigate a specific hazard complaint rather than for a routine workplace inspection, but they do have teeth – the power to issue improvement or prohibition notices, or to prosecute the employer. (See Appendix C at the end of the book for more details.)

Should an inspector refuse to visit your workplace, or should you feel the advice inadequate, you should make a complaint to that person's manager. If you are still not satisfied you can make a written complaint to the HSE Director General at Health and Safety Executive, Rose Court, 2 Southwark Bridge, London SE1 9HS.

While enforcement is limited by the numbers of inspectors – there were job cuts of 20 per cent between 1979 and 1988 – the HSE continues to publish pamphlets about worker health and safety. Managers of small shops, offices and workshops usually do not have sufficient information about workers' rights and safety, so booklets of this type are especially helpful if you work in one of these small companies. In 1994 the HSE published a booklet *New and Expectant Mothers at Work – A Guide for Employers*, which sets out the health and safety rights and the annexes to the Pregnant Workers Directive (which are in Appendix E of this book) with HSE comments. (It costs £6.25 and is available from HSE Books, PO Box 1999, Sudbury, Suffolk CO10 6FS. Tel: 01787 881165.)

Many unions have their own material on the pregnant worker and these pamphlets may be more beneficial to you as they are more realistic about working conditions and the implementation and enforcement of regulations. You can also get the leaflet *Health and Safety at Work: Your Rights in Pregnancy and After Childbirth* by sending a stamped self-addressed envelope and a cheque or postal order for £1.00 to the Maternity Alliance, 45 Beech Street, London EC2P 2LX.

Action Through Your Trade Union

If you are a member of a trade union, your health or safety problems at work should be of concern not only to your own shop steward and branch, but to the union as a whole. In some cases the whole trade union movement through the TUC could support your cause. Not surprisingly, branches vary widely in respect of the amount and type of support you might expect if you suspect some substance or process is impairing your health.

In order to strengthen the role of women within the union movement, the Trades Union Congress (TUC) has instituted an annual women's conference. Every union can send a maximum of twelve delegates whose mandate it is to discuss issues that affect women members both inside and outside the workplace. The conference approved a new TUC charter for women at work which sets out action points on 18 key issues affecting women workers.

In addition to participating in the women's conference, you can take two lines of approach to obtain action:

1. Through the Union's Safety Representative

The Safety Representative has the right under the Safety Representatives and Safety Committees Regulations 1977 to represent workers' concerns, to investigate potential hazards and health and safety complaints by employees, and to inspect the workplace. To underscore the importance of this function, the TUC designated 1989 as 'Inspect and Protect' year. All safety representatives were encouraged to inspect their workplace at least once every three months (the legal minimum) and to report on how effective they were. (See Appendix C for more details about safety representatives. If there is no recognised union see below.)

This avenue should be used if you hope to get the problem dealt with locally and quickly. For example, if you need better fitting protective clothing for yourself and other pregnant workers. An inspector or independent consultant would be called to investigate say, vibration, or dust, or vapour levels at the workplace thought to be harmful to reproductive health. If the inspector affirms the danger, the employer needs to provide the protective clothing.

You should also ask your safety representative to ensure that the employer's risk assessment includes reproductive risks, even if no one is pregnant at the moment.

2. Through Your Union Branch

Use this route if you want to improve prospects of health and safety for workers in general by calling for increases in numbers and powers of the HSE Inspectorate, or banning use of a particular substance or process. You can put a motion to a meeting of your branch requesting the union executive to take the necessary action. A helpful first step would be to check on the union's annual report to see what policy had been so far, so that the wording of your motion might support or extend current policy. Second, ask your branch to explain the aim of this motion to others. It is important to get additional support when it goes to the area meeting. This is crucial as it is there that the decisions on what motions are to be put to the annual conference get made. From there it might, if successful, go to the trades union conference and, if approved by delegates from other unions, would have the weight of all unionised workers behind whatever action is taken.

An example of union action can be seen in the fight over toxic chemicals, specifically the effort to ban weed killers containing 2,4,5-T (one of the constituents of the Vietnam defoliant Agent Orange). Early disquiet about skin troubles, miscarriages, and deformities in the families of agricultural workers who were exposed to 2,4,5-T led the National Union of Agricultural and Allied Workers to campaign to prevent further use of the herbicide. This campaign was actively supported by many other unions as workers in many different industries were affected. Those at risk included local authority workers in parks and gardens, lorry drivers, dockers, process workers, railway and shop workers, amateur gardeners and members of the public.

The Minister of Agriculture refused the unions' call for a ban, and in 1982 the government ignored an EEC recommendation that member nations should not spray 2,4,5-T near public footpaths or near wild fruit which can be picked. Instead the government came up with an unhelpful and unpopular suggestion that more public footpaths be closed. In reaction to this irresponsible decision, the unions mounted a public relations campaign which pointedly showed that the system used to control pesticide use in the UK was haphazard, unsafe and unenforceable. The TUC also passed a resolution calling for complete overhaul of the system. The campaign had the effect of helping to bring about formation of:

- the trade union working group on reproductive hazards
- a farm chemicals sub-committee of the HSE's Agricultural Industry Advisory Committee, to deal with problems of workers using the chemicals
- a trade union farm chemicals working group of concerned scientists, doctors and trade unionists.

In addition, 100 local authorities and many major employers ended their use of 2,4,5-T due to the refusal of many unionised workers to handle it.

A similar pattern of toxicity has become apparent in other countries, with similar concern and action among workers. But there are large stockpiles in Britain and as of 1992, it was still on the Ministry of Agriculture's list of approved herbicides. Of equal concern is the power of the British Agrochemical Association which has not been a friend of workers in general, let alone pregnant workers. The retiring chairman said in May 1981 '...If we give way on 2,4,5-T the unions will go on to campaign against another chemical and then another. We are engaged in a power struggle for control over the industry which we cannot afford to lose.'

Fighting for safety at work has never been easy, but more can be achieved by combined action of the trade union movement than by any individual alone.

Action Where There Is No Recognised Trade Union

Where there is no recognised trade union or you are not a member, you still have the right to be consulted by your employer on health and safety matters. This right is contained in the Health and Safety (Consultation with Employees) Regulations 1996. You can choose whether to be consulted directly or to elect someone to act as a 'representative of employee safety'. A representative of employee safety can make representations to the employer on potential hazards and general health and safety concerns. (See Appendix C for more details.)

ACTION THROUGH YOUR LOCAL AUTHORITY

Many of you work in small businesses and shops where there are no unions to protect you when you believe that your health may be at risk in the workplace. In such a situation, if your workplace is not covered by the Health and Safety Executive, you can seek aid from your local authority. The local authority is responsible for provision of enforcement officers, usually called environmental health officers (EHOs), whose function is to enforce the health and safety laws in about 10,000 premises and to recommend remedial processes where working conditions are unsatisfactory. They operate like HSE inspectors and cover retailing, some warehouses, most offices, hotels and catering, sports, leisure, consumer services and places of worship.

Local government functions in England and Wales (they are slightly different in Scotland and Northern Ireland) depend on the size of the authority. The larger ones, like the county councils, metropolitan districts and London borough councils have responsibility for consumer protection, refuse disposal, police, transport and so on. Non-metropolitan districts are responsible for environmental health, building regulations and so on. These local authorities consist of the officers and of part-time councillors who are directly elected and are increasingly organised along party and political lines.

To get action on an environmental health issue, for example, you could take the following steps:

- Place a well-documented case to one of the councillors for the ward in which you live or work and ask them to take action on it. Support your request, if possible, with a petition from concerned others. (The council offices will have a list of councillors' names, their political affiliations and particular committees on which they are active.)
- Approach the councillor who is chairperson of the committee which would deal with the issue about which you are concerned.

Going as a member of a group can show the councillor that it is an issue on which many workers feel strongly.
- If the council does not take an interest in the issue, bring it to the attention of the local press, radio station and other groups in the area that may strengthen your bargaining power.
- Document how the council's failure to act on the environmental or consumer protection issue will affect local mothers/ workers/residents.

Many people are unaware that it is normally possible for residents to attend council meetings. If you have put forward an issue, attend meetings, so you can follow progress of the issue you have put forward and see which members support or oppose it.

If you want direct action on a more limited issue, such as having an environmental health officer visit your workplace, the council officers will tell you how to go about it.

GETTING HELP FROM OCCUPATIONAL HEALTH SERVICES IN THE WORKPLACE

Getting help through the Occupational Health Service (OHS) staff is a possibility, but do not count on it. In the UK there is no legal requirement for employers to provide an occupational health service, except to a limited degree in dangerous industries such as lead working and radiation. OHS staff usually consist of some or all of the following:

- first-aider
- safety officer
- nurse
- industrial hygienist
- company doctor
- fire officer
- safety engineer.

Apart from the last two, these people may be helpful if you are trying to find out more about a suspected hazard or cause of illness.

In 1985 the International Labour Organisation adopted an Occupational Health Services (OHS) Convention requiring all governments that ratified it to legislate for provision of an OHS free to all employees, but the British HSE recommended further study of the proposal rather than immediate ratification despite the fact that the available evidence indicated that workplace related diseases and injuries were widespread.

In a pilot study in Sheffield GPs found that one out of three patients coming to the surgery, had suffered ill-health as a result of past employment. In the United States, the National Institute for Occupational Safety and Health (NIOSH) estimated that of ten leading work-related diseases and injuries, reproductive disorders (such as miscarriages, sterility and low sperm counts) come sixth in order of frequency. If the UK had the data, a similar pattern would probably be found here. Such figures could be assembled through occupational health services, if all firms had them.

THE POST THATCHER YEARS: THE NEED FOR INCREASED RESEARCH FUNDING

The Thatcher years were hard for workers. Not only was the power of unions curbed, but research into workplace hazards, particularly reproductive hazards, was not viewed as urgent and not given the priority it deserved. Setting research priorities, allocating money to such investigations and designing acceptable studies often ran into political, scientific and economic obstacles, and continue to do so. There is certainly not enough money to fund all worthwhile projects, so it is important to make a good case for studying occupational reproductive hazards.

Clusters of workplace-related harm always need to be investigated carefully. In the past, analyses of clusters of unusual abnormalities confirmed their being related to pharmacological, workplace and environmental hazards. The dangers of properties of methyl mercury, rubella, thalidomide, and DES (diethylstilbestrol) were documented by this route. Clusters of miscarriages and birth defects among VDU operators have been the impetus for studies in several countries investigating the connection between reproductive health and VDU use. Though there is no solid evidence linking VDU use to reproductive harm, two studies have shown a link between miscarriages and working more than 20 hours a week at a video terminal. Whether these results will hold up in further studies, or whether some aspect of the machine, stress or some characteristic of the workers is responsible for the findings is not yet known. Additional research should provide at least some of the answers.

BARRIERS TO PROGRESS

If you do decide to try to obtain a healthier workplace for yourself when you are pregnant, you must determine what hurdles you have to overcome and decide how to meet them. One of the first steps for many women is to develop the confidence and skills needed to win over the opposition. Whether you choose the route of individual action or participate in group endeavours, you may need some assertiveness training before you can achieve your goal successfully and without undue stress.

Many of you believe strongly in gender equality. But when you start to take a forceful stand, you begin to feel uncomfortable. You realise that your deeply held views about equality cover up images of sex roles that you learned as young girls. You are not aware that these early images still exist deep inside yourselves until you begin to behave in assertive ways. 'Will people feel I'm unfeminine' – or insensitive, or overbearing, ruthless and hostile? All these stereotypes come to the forefront and the fledgling asserter begins to feel guilty. Nonsense! None of these characteristics has anything to do with assertiveness. Remember that being feminine, loving, kind, understanding and sensitive is not to be equated with giving up your rights. Ann Dickson's book *A Woman in Your Own Right* (Dickson 1988) offers many good pointers and can function as a do-it-yourself guide. Keep re-reading it when you slip back into old feelings of inadequacy and guilt.

One area in which assertiveness training is especially helpful is in dealing with sexual harassment. Pregnant workers suffer from sexual harassment as well as non-pregnant women and it is widespread in both the US and the UK. Men used to believe it was their right to be 'touchy, feely' or make insulting comments as a joke, and some refuse to change. Now women are speaking up and asserting their right not to be teased or molested. In a British national opinion poll for *Independent on Sunday*, one in six women replied that they had experienced sexual harassment at their factory, shop or office. Libraries and bookshops carry a selection of books on assertiveness training. Here are some of the basic premises that *A Woman in Your Own Right* and other books on assertiveness training present.

Assertiveness

What does being assertive mean?

- Deciding what you want and saying so specifically, directly and calmly.
- Sticking to your position. Repeating it as often as necessary.
- Preventing the person or persons you are confronting from undermining your assertiveness by intimidating you into feeling hostile, guilty or inadequate. If they make you angry, express that feeling directly, but calmly.

What are your rights?

- The right to state your own needs and set your own priorities.
- The right to be treated with respect and dignity.
- The right to be considered an intelligent and competent human being.
- The right to express your feelings, opinions and values.
- The right to make decisions for yourself.
- The right to make your own mistakes.
- The right not to be subjected to undue persuasion, bullying or use of guilt to get you to change your course of action.
- The right to say you do not understand and require additional information and time to come to a decision.
- The right to say no or change your mind.
- The right to ask, organise and fight for what you want and believe you are entitled to.

In addition to lack of assertiveness and undervaluing your own ability, there are additional barriers that have to be overcome before you can be successful:

- lack of factual information
- being unable to pinpoint the problems
- not knowing who to go to in order to get the problem fixed
- believing that nothing will change anyway
- overestimating the ability of scientists to know the answer
- fearing harassment
- having promotion opportunities blocked or being fired.

Information and resources in this chapter and previous ones should provide you with the tools you need to overcome these barriers. You can then decide what efforts you wish to undertake in your own workplace and what regulations or legislation to seek from the government in order to reduce the reproductive risks in the workplace and compensate for prior harm. The appendix at the end of this chapter provides additional information for those of you who might like to investigate getting more involved in the fight for reproductive health in the workplace.

Despite the Conservative party's anti-worker stance, the EU Work Related Directives have enhanced worker protection. While the UK has fought against more generous legislation, UK workers are better off than they would have been if the Thatcher and Major governments had had their way completely. It remains to be seen whether the Labour government's policy of a 'Fair Deal at Work' will add anything to workplace protection, but at least signing up to the Social Chapter will bring some improvements in workers' rights.

THE FUTURE

Public concern voiced by women in the workplace and the community has raised society's awareness of occupational and environmental reproductive hazards. Awareness that the basic disagreements revolve around questions such as what is considered evidence and what policy decisions should be implemented is less widespread. Safety versus profits too often clash, but this bottom line is covered with rhetoric largely cloaked in scientific terms. When millions of pounds are at stake, companies become defensive, callous and insensitive. The pinnacle of insensitivity was achieved by representatives of the Louisiana chemical industry who defended what they considered to be their exemplary environmental record. When asked to respond to the assertion that exposure to vinyl chloride from the petrochemical plants caused the high miscarriage rate found in those living nearby, they answered that there is as little proof that chemicals cause spontaneous abortions as there is that 'screwing' too much is the cause. The head of the British Agrochemical Association came in a close second with his solution of closing public footpaths near wild berry patches rather than switching to a safer pesticide.

With such enemies, we need a lot of friends!

We Shall Overcome

Yes, we need a lot of friends! But we already have lots of friends – for the best friends we have are other women. Women in cooperation and coalition with each other and with concerned men will overcome resistance to achieving a pregnancy friendly workplace. The road ahead is still a long and hard one, but it should be easier than it has in the past. As we have seen, some progress has been made. The issue of reproductive hazards of the workplace has been raised nationally in the media many times. The EU has addressed the issue, both directly and indirectly, with a series of work related directives, and these have been translated into UK regulations. The image of woman as sole vehicle of reproductive damage has been, if not demolished, at least questioned. Males themselves are beginning to realise that their reproductive functioning also can be damaged by workplace toxins and processes. Childcare and family leave agendas are now being widely discussed.

Women concerned about environmental and occupational health and reproductive hazards need to participate actively in all stages and facets of decision making on workplace hazards and risk. It is to be hoped that in the future women will be treated as equal partners in evaluating workplace risks and their viewpoints taken more seriously. Only then can women, together with scientists, government officials, business leaders, healthcare workers and the British public achieve a reproductive friendly workplace for all.

CHAPTER 5 APPENDIX 1: ASSERTIVENESS TRAINING SKILLS

Role-playing is the most useful technique in learning to become comfortable about being assertive. Ask a good friend or co-worker to role play with you and then give each other feedback. Simulate a real situation involving pregnant workers and management and you'll be surprised that after the first few awkward minutes, you will be feeling the part.

Start with an easy case first and then build up to the more difficult one. For example, suppose that you want to bring your grievances about suspected reproductive hazards in the workplace to the attention of senior management. You think you will be able to get a more sympathetic hearing if you can get some supervisory personnel on your side. You know of one person that you might be able to convince. Start your role-playing with this person in mind. Take the part of the assertive advocate and have a co-worker who knows the supervisor in question take the supervisor's role.

In addition to being clear and specific about your demands, you must make sure that your body language does not cancel out your assertive verbal message. One common mistake in the early stages of assertiveness training is to concentrate so hard on what you are saying and how you say it that you forget to think about your tone of voice, facial expression and posture. Once you are really *feeling* assertive, the assertive body language will follow naturally. But meanwhile you may have to unlearn some counter-productive body habits.

What Not To Do

- Do not slouch, shift your weight from one foot to another, keep your head cocked to one side, fidget or shuffle your feet.
- Do not stay too far away from the people you want to influence or let them tower over you.
- Do not avoid eye contact or convey either hostility or timidity.
- Do not clench your jaw, tighten your lips or thrust your shoulders forward.
- Do not paste a false smile on your face or grimace.
- Do not speak too loudly, too softly, mumble or whine, or use a simpering little girl's voice or an apologetic or sarcastic tone.

What To Do

- In role-play, ask an observer or your role partner to point out if you make any of the above mistakes. They can usually be corrected by readjusting your voice or body movement only slightly.
- Take several deep breaths to calm yourself before approaching a confrontation.
- Sit or stand in a well-balanced position, close enough to catch the attention of your audience.
- Speak slowly and distinctly. Modulate your voice varying the pitch from the high to low registers of your vocal chords.
- Relax your shoulders and keep your arms at your sides.
- Dress in a way that makes you feel good about yourself.

Whether you read a book on assertiveness training or take a course, it is a good idea to do some role-playing to make you feel more secure and to strengthen your approach. You can even video your presentation and play it back.

There are a number of good references on assertiveness training and effectiveness:

A Woman in Your Own Right, Anne Dickson. London: Quartet Books, updated and revised 1988.
Assert Yourself, Gael Lindenfeld. London: Thorsons, 1988.
How to Survive as a Working Mother, Lesley Garner. Harmondsworth: Penguin, 1982.
Self-Assertion for Women, Pamela Butler. London: Harper and Row, 1976.
Women's Rights: A Practical Guide, Anna Coote and Tess Gill. Harmondsworth: Penguin, 1977.

Most adult education programmes offer classes in assertiveness training and unions sometimes include it as part of their education and organising programmes. Check with your local union branch and local educational institutions.

CHAPTER 5 APPENDIX 2: TAKING ACTION

How to Get Action Taken Through Parliament

Obtaining action through Parliament would require enacting a new law, or a change in existing law. Unless the legislation you want is close to the interests of the current government, so that they would take up the matter and incorporate it in their own programme, the best route is via a Private Members Bill; Bills for which a relatively small amount of parliamentary time is allotted. There is great competition amongst MPs of all parties to initiate legislation under this heading and the practice is to draw lots for the time allowed, so it is implicitly a chancy process, but if an MP is interested, he/she will follow up failure in one session with a further application for the next occasion. (This was the case with the Bill presented by David Alton to restrict the period after conception during which abortion might be performed, and it should be noted that it was not part of his own party's policy.)

First Steps (as an individual or as member of a group)

- Prepare the evidence of the abuse of the current system, and the laws under which it operates. Try to get some legal help in this to make the issue as clear-cut as possible. MPs will be more willing to get involved if they see that they do not have to do all the groundwork.
- Make an appointment with the MP for your constituency (your local library will have a list), either at the House of Commons or at his/her office in the area in which you live. Alternatively you might approach a member interested in issues affecting working women, health and safety, industrial illness and so on, or known to have had personal experience of such working conditions.
- Try to persuade the member of the need for legislation, its importance for considerable numbers of workers, and of the effects for the future if such legislation is not enacted.
- If the MP is not willing to put forward a Private Members Bill, don't give up. Ask him/her to suggest another MP you might approach; to ask Parliament for an investigation to be set up to find out how widespread are the conditions or problems you are complaining of; or to ask a question in the House of Commons so phrased that the answer could be used to pressure employers about the issue.

How to Get Action Through a Political Party

Remember that this will not be a quick way to achieve the desired action. Remember too that political parties are not one-issue organisations. They are coalitions of people who have come together to seek power in order to achieve a wide variety of goals. They are places for horse-trading and compromise. However, joining a political party does give you a chance to influence the policies to which it is committed in its manifesto and the people it chooses to stand as its candidates at the next election. To do this you will have to attend meetings, join the right committees and persuade other people to agree with your position. Even then, you might not succeed. There is a great difference between what the members of a party want and what the MPs they elect actually give them.

The address of your local branch can be found in the telephone directory, and once you are a member you have the right to go to branch meetings, vote and put resolutions. The more active you are at this level of the party, the more chance you have of influencing what the rest of the party does – you can start resolutions which go right up to the national party conference. Selection of these would probably be decided by the general management committee, and voting for the member of this committee is an important aspect of ensuring that people who might be expected to support your interest are elected to it. An executive committee of this body will be likely to make the day-to-day decisions such as short-listing candidates to represent the party as Member of Parliament for that constituency, to be interviewed by the selection committee.

Being an effective member of a political party may take a good deal of time, but it is another way of publicising and perhaps getting action on the important matters that you want to see changed.

CHAPTER 5 APPENDIX 3: ORGANISING TECHNIQUES – A CHECKLIST FOR SUCCESSFUL DO-IT-YOURSELF ACTION

This is primarily for non-unionised workplaces with hostile managements or those in which unions are not sufficiently interested in reproductive hazards.

Again, organise together with men if possible. Times are changing and men are becoming more aware of the reproductive damage they may suffer. In addition, many men now have working wives and workplace protection for pregnant women is becoming more of a personal issue for them. In some cases, such as that of a pregnant middle manager, there may not be enough staff at the same level to form an action group. In these cases a small group of workers can do effective research and present their concerns. Sometimes employers are ignorant rather than hostile and can be persuaded with well-documented material presented in a non-hostile fashion by valued senior employees. With the implementation of the European Union Directive and Britain joining the Social Charter, managements may be entering an era of increased awareness of the importance of workplace health and safety.

Choosing the Issue

Gather information and identify which problem areas require correction. Survey the workers regarding their concerns and ascertain that the suspected workplace reproductive hazard is considered serious enough for the women to be willing to act.

Be sure that you have some evidence to back your demands. This can either be from a scientific study or from workers' reports regarding a suspiciously high number of reproductive health problems in your own workplace or others. It is essential that you have some feasible solutions in mind and that the women believe that they can do something about correcting the problem.

Choosing the Type of Action

Choose the type of action that the women feel comfortable in undertaking and maintaining.

Attempt to judge how hard your employers will fight back and what compromise they might settle for. Decide whether or not you are willing to settle for half a loaf.

Try to figure out what price you are likely to pay for your action if it is not successful. Let the women decide whether the chance of success is worth the penalty of failure.

Preparing for the Action

Make sure you have provided enough information and involved enough women in planning the action. Women need to believe that they can win and be reassured that they can effectively perform the disagreeable tasks that are part of the process of exerting pressure. Delegate work and motivate the fainthearted. Share the decision making. Individual strengths can be utilised and individual weaknesses counterbalanced in a group effort. People are more likely to feel stronger about an action if they have a part in designing the strategy.

Maximise the Threat Before Initiating Action

Be sure that your demands are clear and feasible and that you have ranked them in terms of importance. If in the bargaining process some of the demands have to be given up, there is then existing agreement on which these are to be.

Organise the Action Thoroughly
It is essential that each woman knows what she has to do and is willing and able to do it. If you cannot deliver on your threat your future bargaining power is strongly diminished.

Plan to Publicise Your Victory in Order to Strengthen Your Position in the Future
Have press releases written. Arrange to appear on TV and radio. Make sure personnel in other divisions of your company learn about your successful action. Inform other unions and organisations and any local community action groups and women's groups that might be your natural allies.

You may not want to become an active organiser yourself, but being aware of the essential techniques enables you to make helpful recommendations to those who have undertaken these tasks and to warn against positions that are ill thought out or unworkable. An informed work force improves the likelihood of success. Suggestions for influencing legislation and specific organising tactics are listed below.

Choosing the Tactics

Your choice of tactics partly depends on the public relations and goodwill costs involved in fixing the hazard as well as the monetary cost of cleaning up the workplace. By a clever use of publicity or the threat of going public, you can sometimes raise the cost of the non-monetary factors high enough for the employer to feel it is cheaper to expedite the safety measures. Here are some approaches you can try that other groups of workers have used successfully.

Group Education
Hold these educational sessions at lunch time so that workers from different parts of the company who are upset about the suspected reproductive hazards can come together and receive the same information. This reduces the misunderstandings that sometimes happen when different people provide information at different times.

Threaten Publicity

Prepare a flyer publicising your grievances. Show the management the flyer you are planning to distribute before you actually circulate it. Tell the management that you will give them a chance to institute remedial action within a specific time frame before you implement your plan.

Use the Media

Local newspapers, TV and radio stations are interested in occupational and environmental hazard stories, especially if you can make them sound dramatic. Work-related fertility problems, miscarriages and birth defects find a ready audience.

Conduct Well-Directed Informational Picketing

Address it to the workers, to the public, to the company and to politicians and civic leaders. Sometimes merely making information public and visible becomes a catalyst for action.

Innovative Actions

Some suggestions for you to use if you do not have a union or as a union sponsored activity.

A Nurse-Out: Encourage all workers, male and female, who suspect that they are exposed to reproductive health hazards in the workplace to visit the company occupational health nurse with complaints during working hours. Substances that are capable of inflicting reproductive damage usually have other harmful effects as well.

A Lunch-Out: When the weather is warm, have lunch together on the company grounds at a site visible to the public. Bring large signs stating your grievances or fly helium balloons printed with an appropriate slogan. Keep this up until the company is ready to make the workplace safer.

A Warm-Out or Cold-Out: If the workplace is too hot or too cold, go to work but only stay in those parts of the workplace that are comfortable. Refuse to work where the temperature is dangerous to your health or your pregnancy.

Be Your Own Reproductive Health Detective

Try to trace the specific source of the problem. Enlist the support of both male and female workers. Ask them to write down any part of the work process, location or time of day when they have suspicions that toxic substances to the reproductive system are accumulating. Ask them to be as detailed as possible, describing odours, vapours, colour changes, dust particles etc. and any immediate or delayed symptoms they suffer.

Appendix A

Occupational Information for your Midwife or General Practitioner

Provide specific information about your current job and fill in additional forms for your past occupations. Include summer, temporary and part-time work.

A. Job description.

1) Describe exactly what tasks you perform ....
2) Describe a typical work shift in detail ....
3) Describe the type of workplace and work station ....
4) Describe any unusual or overtime tasks ....

B. How long have you been/were you in the job?

....
....
....

C. List any new or changed processes at work.

....
....
....

D. List chemical, physical, biological and psychological stresses at work.

1) What are the chemical names of the substances you are exposed to at work?

2)Use your rights under COSHH regulations.

....

....

....

E. Estimate the extent of your exposure.

(e.g., 'My clothes are covered with a fine film of dust an hour after I start work.')

....

....

....

F. Provide detailed information about eating, drinking, and smoking in the workplace.

(i.e., what you and others do, for how long you do them, where you do them, and what work processes are going on concurrently.)

....

....

....

G. If you wash or shower at work, describe the facilities and what you do with your clothing .

....

....

....

H. If you wear any protective clothing or hearing protectors, describe the fit, how often you wear them and how comfortable they are.

....

....

....

I. Give details of what you have found out from your employer or union regarding whether protective engineering systems and devices such as exhaust and ventilation systems are

installed, whether they are functioning and whether they are adequate.

....

....

....

J. Monitor your symptoms and compare them with co-workers.

1) How soon after you get to work do your symptoms start? ....

2) How soon after you get home do your symptoms stop? ....

3) Do your symptoms feel worse when a special process is being performed? If so, what is that process? ....

4) Is there a pattern of symptoms among your co-workers? ....

5) Are there other factors not connected to work that might solely or in combination with workplace exposures be causing your symptoms?

(a) Someone else in the household may be bringing home a hazardous substance on work clothes ....

(b) Substances in the home such as art material, cleaning fluid, pesticide spray ....

(c) Location of home near factory, incinerator, rubbish dump or contaminated source of water ....

(d) Use of new cleaning products, soaps, cosmetics, clothing........

(e) A hobby that requires the use of hazardous material ....

(f) Smoking cigarettes, cigars or pipes or heavy use of alcohol in the past or present ....

(g) Previous change of residence because of a health problem

....

Source: excerpted from 'Recognizing and Preventing Hazards in the Workplace', Barry S. Levy, *Consultant*, Nov. 1983: 63–78; and 'The Occupational History', Occupational Health Committee, *Annals of Internal Medicine*, Nov. 1983, Vol. 99, No. 8, 643–644.

Appendix B

Agents and Substances Reviewed for Reproductive Health Effects by the Congressional Office of Technology Assessment

AGENTS AND SUBSTANCES

Chemicals

Agricultural chemicals:
- carbaryl
- dibromochloropropane (DBCP)
- DDT
- kepone (chlordecone)
- 2,4,5-T, dioxin (TCDD), and Agent Orange 2,4-d

Polyhalogenated biphenyls:
- polybrominated biphenyls (PBBs)
- polychlorinated biphenyls (PCBs)

Organic solvents:
- carbon disulphide
- styrene
- benzene
- carbon tetrachloride
- trichlorethylene

Anaesthetic agents:
- epichlorohydrin
- ethylene dibromide (EDB)
- ethylene oxide (EtO)
- formaldehyde

Rubber manufacturing:
- 1,3 butadiene
- chloroprene
- ethylene thiourea

Vinyl halides:
- vinyl chloride

Hormones
Undefined industrial exposures:
agricultural work
laboratory work
oil, chemical and atomic work
pulp and paper work
textile work

Physical Factors

Ionising radiation:
X-rays
gamma rays
Non-Ionising radiation:
ultraviolet radiation
visible light
infrared radiation
radiofrequency/microwave
laser
ultrasound
video display terminals
magnetic field
Hyperbaric/hypobaric environments:
hot environments
cold environments
noise
vibration

Stress

Metals
Lead
Boron
Manganese
Mercury
Cadmium
Arsenic
Antimony

Biological agents:
Rubella
Cytomegalovirus
Hepatitis-B

Other infectious agents
Recombinant DNA

Source: US Congress, Office of Technology Assessment, *Reproductive Health Hazards in the Workplace*, Washington, US Government Printing Office, December 1985, p.7.

SOME SUBSTANCES THOUGHT TO CAUSE ADVERSE REPRODUCTIVE HEALTH EFFECTS IN ANIMALS OR HUMANS DUE TO OCCUPATIONAL EXPOSURE

Lead

Exposure	*Female*	*Male*
Prior to conception	• possible chromosome aberration (h) • menstrual disorders (h)	• possible chromosome aberration (h) • sperm abnormality (h) • degeneration of testes (h) • decreased sex drive (h)
At conception		• sperm abnormality (h)
During pregnancy	• miscarriages (h) • stillbirths (h) • malformations (a)	
On newborn	• lead in breastmilk (h)	
On child (through clothing)	• lead poisoning (h) • hyperactive (h) • brain damage (h)	• lead poisoning (h) • hyperactive (h) • brain damage (h)

Anaesthetic Gases

During pregnancy	• miscarriages and birth defects (h) (female and male exposure)	

Vinyl and Polyvinyl Chlorides

Prior to conception	• chromosome aberrations (h)	• chromosome aberrations (h) • mutations in genetic material in sperm (a)

Exposure	*Female*	*Male*
During pregnancy	• cancer in offspring	• miscarriages and stillbirths in partners (h)
Benzene, Toluene, Xylene		
Prior to conception	• damaged chromosomes (h) • menstrual disorders (h)	• damaged chromosomes (h)
During pregnancy	• aplastic anaemia in woman (h) • birth defects (h)	
Chlorinated Hydrocarbons		
Prior to conception	• chromosome aberrations (h)	• chromosome aberrations (h) • infertility (a)
During pregnancy	• liver damage to foetus (a) • in breast milk (h) • miscarriages (h) • cancer in offspring (a) • spinal birth defects (h) • stillbirths (a)	
Oestrogen		
Prior to conception	• effect on flow, frequency of menstrual cycle (h)	• sore and enlarged breasts (h) • impotence (h)
On child (through clothing)	• sore and enlarged breasts of prepubescent (h)	• sore and enlarged breasts of prepubescent (h)
Ionising radiation		
Prior to conception	• mutations in genetic material (a) • reduced fertility (h)	• mutations in genetic material (a) • reduced fertility (h)

Exposure	*Female*	*Male*
During pregnancy	• leukaemia and other cancers in offspring (h) • cataracts and eye defects in offspring (a) • small heads and brains in offspring (a) • learning disabilities and retarded growth in offspring (h)	
Pesticides		
At conception	• prevent conception (a)	
During pregnancy	• stillbirths (a) • miscarriages (a) • abnormal offspring (a) • leukaemia in pregnant mother and offspring (a)	
On newborn	• pesticides in breastmilk (h)	
Carbon Monoxide		
During pregnancy	• smaller size of newborn and higher chance of dying soon after birth (a) • stillbirths, cerebral palsy, learning disabilities in offspring (h)	
Carbon Disulphide		
Prior to conception	• irregular menstruation (h) • extreme bleeding (h)	• decreased sex drive (h) • sperm abnormalities(h) • impotence (h)

Exposure	*Female*	*Male*
At conception	• decreased fertility (h)	
During pregnancy	• miscarriages (h)	
PCBs		
At conception	• reduced ability to become pregnant (a)	
During pregnancy	• small babies with PCBs in tissue (a)	
	• stillbirths (h)	
	• babies born with discoloured skin which then fades (h)	
On newborn	• PCBs in breast-milk (h)	

Key: (h) At least one study on humans showed this effect.
(a) At least one study on animals showed this effect.

Sources: , Barlow, Susan M. and Frank M. Sullivan *Reproductive Hazards of Industrial Chemicals: An Evaluation of Animal and Human Data*, London: Academic Press, 1982; Hricko, A. and Melanie Brunt, *Working for Your Life: A Woman's Guide to Job Health Hazards*, Berkeley, Labor Occupational Health Program, University of California, 1976; US Congress, Office of Technology Assessment, *Reproductive Health Hazards in the Workplace*, Washington, US Government Printing Office, 1985.

Appendix C

UK Workplace Protection – Acts and Regulations

The laws which protect health and safety are found in a number of different acts and regulations. The main act is the Health and Safety at Work Act 1974 (H&SAWA) which lays down general principles. The H&SAWA also gives the government power to bring in regulations which are more specific, covering either particular types of hazard or particular ways of dealing with risks.

Many of the regulations have been introduced because of EU laws known as directives (see Appendix E). When a directive has been passed in Europe, the Health and Safety Commission (see below) puts forward proposals for new regulations, and interested groups can comment on the proposals. The regulations are laid before Parliament for a vote and if accepted they become law. The Health and Safety Executive (see below) is responsible for enforcing them. It also produces approved codes of practice and guidance notes to explain the regulations to employers.

The most important regulations are the Management of Health and Safety at Work Regulations 1992 which set out the process of risk assessment and the steps that follow, including those that apply to reproductive risks.

THE HEALTH AND SAFETY AT WORK ACT 1974 (H&SAWA)

This Act sets out employers' responsibilities and workers' rights in very general terms. Employers are obliged 'to ensure, so far as is reasonably practicable' their employees' health, safety and welfare at work. Before the Act was introduced, workplace protection depended on the *type* of workplace or process, but this Act applies to all workplaces, employers and employees. An employer who breaks the Act can be prosecuted by the Health and Safety Executive (see below), although the HSE inspectors are more likely to make suggestions, issue an improvement notice or (if there is a risk of serious personal injury) issue a prohibition notice.

The Act also gives the government power to introduce health and safety regulations on more specific issues. Examples are the MHSW Regulations, the COSHH Regulations, and other regulations introduced to implement the EU directives.

REGULATIONS ABOUT GENERAL HEALTH AND SAFETY RIGHTS

The Management of Health and Safety Regulations 1992 (MHSW)

The basic protection for pregnant workers is found in these regulations which were updated in 1994 (*Management of Health and Safety at Work (Amendment) Regulations*) specifically to include female reproductive risks. The MHSW regulations require every employer to make a 'suitable and sufficient assessment' of the risks to the health and safety of his employees and others who might be affected in order to identify what measures he/she needs to take to comply with any relevant statutory provisions. He/she must also provide employees with comprehensible and relevant information on any risks identified by the assessment and the preventative and protective measures taken.

The regulations apply to all employees but you are specifically protected if you are a 'new or expectant mother', which means that

- you are pregnant, or
- you are breastfeeding, or
- you have given birth (including stillbirth) within the last six months.

To get the full protection you must

- tell your employer in writing that you are pregnant or breastfeeding, or that you have given birth in the last six months
- if your employer asks in writing for proof that you are pregnant, show him/her your certificate of pregnancy from your doctor or midwife.

Your employer must:

1. Carry out a 'risk assessment' of any processes, working conditions, physical, chemical and biological agents which could jeopardise your health or safety or that of your child while you are pregnant, breastfeeding, or if you have given birth within the previous six months.
2. If a risk is found, do all that is reasonable to remove it or prevent your exposure to it, for example by complying with any relevant regulations covering that type of hazard.
3. Give you information on the risk and the action that has been taken.
4. If the risk remains, temporarily alter your working conditions or hours of work, if this is reasonable and if it avoids the risk.
5. If the risk can't be avoided, offer you suitable alternative work (on terms and conditions which are not substantially less favourable than your original job).
6. If there is no suitable alternative work available, suspend you on full pay (i.e. give you paid leave) for as long as is necessary to avoid the risk.

All employers who employ any women of childbearing age are legally required to take steps 1 to 3, *even if none of the employees are pregnant at the moment*. This means that unions or safety representatives should routinely ask for the employer's normal risk assessment to include risks to new or expectant mothers, rather than leaving it up to an individual woman when she becomes pregnant. Steps 4 to 6 do not apply until a woman is actually pregnant and has notified the employer in writing and produced a medical certificate confirming pregnancy if requested.

The regulations tell employers to include in their risk assessment the working conditions and biological and chemical agents listed in the annexes to the Pregnant Workers Directive. These annexes are a very useful starting point for thinking about risks, and are set out in Appendix E on pages 228–9.

Remember that a pregnant woman cannot be sacked because she can't do the same work as before her pregnancy. If the employer cannot make her job safe or offer her suitable alternative work then she must be suspended on full pay (not made to take sick leave).

Night work: The regulations also state that where a new or expectant mother who works at night has a medical certificate stating that night work could affect her health or safety, the employer must either offer her suitable alternative daytime work (if there is any available), or suspend her from work (give her paid leave).

The Workplace (Health, Safety and Welfare) Regulations 1992

These regulations state that as of 1 January 1996, all employers must provide 'suitable facilities' for a pregnant woman or nursing mother to rest. The Code of Practice states that rest facilities 'should be conveniently situated in relation to sanitary facilities and, where necessary, include the facility to lie down' (paragraph 237).

The Safety Representatives and Safety Committees Regulations 1977

These regulations give trade unions the right to elect their own safety representatives. If at least two safety representatives request it, the employer has to set up a safety committee. The function of a safety committee is to keep the health and safety measures at the workplace under review.

It is estimated that there are over 200,000 safety representatives in the UK. They have three types of rights:

1. To *represent* the employees in health and safety consultations with the employer, and to receive health and safety information on their behalf.
2. To *investigate* potential hazards and the causes of accidents, and to investigate health and safety complaints by individual employees.

3. To *inspect* the workplace:
   a) three-monthly routine inspections, planned in advance, which are often carried out jointly with a member of management
   b) inspections based on workplace changes or new information. These can be conducted on a day to day basis
   c) inspections carried out in response to the discovery of an occupational hazard or disease, a dangerous occurrence or accident
   d) inspection of relevant documents.

Unfortunately these regulations only apply where there is a trade union recognised by the employer. Non-unionised workplaces are covered by the Health and Safety (Consultation with Employees) Regulations 1996 (see below).

The Health and Safety (Consultation with Employees) Regulations 1996

These regulations apply where there is no union or it is not recognised, so there are no official safety representatives. The employer must consult employees 'on matters relating to their health and safety', and in particular on the introduction of any measure or new technology which may affect their health and safety. The employees can elect one or more of their number to represent them in these consultations (they are then known as 'representatives of employee safety'), or the employer can consult the employees directly. He must give the employees or their representatives adequate information to enable them to 'participate fully and effectively'.

The functions of the representatives of employee safety are:

1. to make representations to the employer on potential hazards and dangerous occurrences at the workplace
2. to make representations to the employer on general matters affecting health and safety
3. to represent the employees in consultations with an inspector.

Representatives of employee safety have the same protection against being victimised or dismissed as union safety reps (Employment Rights Act sections 44 and 100).

REGULATIONS ABOUT PARTICULAR HAZARDS

The Provision and Use of Work Equipment Regulations 1992

These regulations require employers to ensure that all work equipment (including machinery, appliances, apparatuses and tools) is suitable for the purpose for which it is used, having regard to any health and safety risks in the workplace including those from the equipment itself. Employers must ensure that employees using the equipment are adequately trained.

The Personal Protective Equipment at Work Regulations 1992

These regulations require employers to provide their employees with suitable 'personal protective equipment' ('PPE'), i.e. equipment which is worn or held to protect a person against a risk to his/her health and safety. They must keep it in good working order and repair, and train or inform employees how to use it correctly. The first step, however, is to control the risk by other means.

Significantly for women whose shape is changing during pregnancy, the PPE is not considered suitable unless a) it takes account of the ergonomic requirements and the state of health of the person who may wear it (regulation 4(1)(b)), and b) it is capable of fitting the wearer correctly (regulation 4(1)(c)).

The Manual Handling Operations Regulations 1992

These regulations require employers to *avoid* the need for employees to undertake any manual handling operations (i.e. lifting, putting down, pushing, pulling, carrying or moving loads by hand or bodily force) involving a risk of their being injured. Where it is not reasonably practicable to avoid this, employers must:

1. make a suitable and sufficient risk assessment of the manual handling operations
2. take appropriate steps to reduce the risk of injury to the lowest level reasonably practicable
3. give employees information on the weight of each load and (if appropriate) its heaviest side.

As part of the risk assessment the employer *must* consider whether the job creates 'a hazard to those who might reasonably be considered to be pregnant'.

The Guidance Notes to these regulations add: 'Allowance should be made for pregnancy where the employer could reasonably be expected to be aware of it, i.e. where the pregnancy is visibly apparent or the employee has informed her employer that she is pregnant. Pregnancy has significant implications for the risk of manual handling injury. Hormonal changes can affect the ligaments, increasing the susceptibility to injury; and postural problems may increase as the pregnancy progresses. Particular care should be taken for women who may handle loads during the three months following a return to work after childbirth' (note 99).

The Health and Safety (Display Screen Equipment) Regulations 1992

These regulations require employers to assess the health and safety risks from VDU workstations and to reduce any risks to the lowest extent reasonably practicable. The employer must also ensure that the workstations comply with minimum health and safety requirements as to the screen, keyboard, desk, space, lighting, glare, noise, heat, radiation and humidity, which are set out in the regulations. Workers using VDUs are entitled to periodic breaks or changes of activity to reduce their VDU workload. Employers must provide a free eye test if an employee requests it and free 'special' corrective appliances, i.e. glasses to correct vision defects at the viewing distance specifically used for the VDU work.

The Guidance Notes to these regulations state that there is no need for VDU users to be given anti-radiation screens because so little radiation is emitted from current equipment. Annex B sets out the Health and Safety Executive's position on VDUs and pregnancy, which is based on information from the National Radiological Protection Board:

> There has been considerable public concern about reports of higher levels of miscarriage and birth defects among some groups of visual display unit (VDU) workers in particular due to electromagnetic radiation. Many scientific studies have been carried out, but taken as a whole their results do not show any link between miscarriages

or birth defects and working with VDUs. Research and reviews of the scientific evidence will continue to be undertaken.

In the light of the scientific evidence pregnant women do not need to stop work with VDUs. However, to avoid problems caused by stress and anxiety, women who are pregnant or planning children and worried about working with VDUs should be given the opportunity to discuss their concerns with someone adequately informed of current authoritative scientific information and advice.

The Control of Substances Hazardous to Health (COSHH) Regulations 1994

These regulations require the employer to prevent, or if that is impossible, to control adequately, exposure to all substances such as chemicals, micro-organisms and dust which can be hazardous to the health of workers. The substances include carcinogens under the Carcinogens Directive, and biological agents under the Biological Agents Directive.

The employer must do the following:

1. Assess the risk to health in the workplace and devise precautions needed. This assessment must be reviewed and if necessary updated, if any of the circumstances of the work should change or if it becomes apparent that the original assessment is no longer valid. Workers or their representatives must be informed of the result.
2. Introduce measures to prevent exposure to the hazard, or, where that is not reasonably practicable, to control it adequately.
3. Ensure that proper procedures and controls are used and that equipment is properly maintained and used.
4. If needed, monitor the exposure of the workers and carry out surveillance of their health.
5. Provide workers access to monitoring and medical surveillance records.
6. Educate and train workers about the precautions that they should take to avoid workplace risks.

Taking the COSSH regulations together with the MHSW regulations (see above), employers have to assess whether the substance under consideration could be harmful to the human reproductive process. A further requirement of COSSH is that employees must be informed of results of any monitoring and collective results of any health surveillance, as well as risks involved in the work and precautions to be taken. So under the regulations you have the right to know whether substances that you are using are at levels that could be harmful to your pregnancy.

The Control of Lead at Work Regulations 1980

The Approved Code of Practice requires that 'women of reproductive capacity' be suspended from work at a lower level of blood lead concentration than others (40 microgrammes/100ml). A woman should notify her employer as soon as possible when she becomes pregnant and a pregnant woman should be suspended automatically from work involving exposure to lead. There is currently (1997) an HSC proposal to reduce the blood lead suspension level to 30 microgrammes/100ml for women of reproductive capacity.

The Ionising Radiation Regulations 1985

These regulations set a lower permitted dose level of ionising radiation for 'women of reproductive capacity' (13 mSv in any consecutive three-month interval) than others, and lower still for pregnant women (10 mSv during the declared term of pregnancy).

Other Useful Legislation

The Access to Medical Reports Act 1988 provides limited access to medical reports prepared for employment purposes.

The Local Government Access to Information Act 1985 provides access to council documents and meetings. See Community Rights Project under the organisations section.

The Health and Safety Commission and Executive

The Health and Safety Commission (HSC) is the overall government body responsible for the development and enforcement of workplace law on health and safety. Its primary functions are:

- to secure the health, safety and welfare of people at work
- to protect the public from risks arising from work activities.

It is made up of eight representatives of trade unions, employers and local authorities. The HSC has an executive (enforcement) arm called the Health and Safety Executive (HSE) which works through twenty area offices and is composed of:

- inspectors who visit and review work activities, giving expert advice and guidance, and where necessary, issuing enforcement notices and initiating prosecution.
- the Employment Medical Advisory Service whose members are doctors or nurses.

The HSE enforces the Health and Safety at Work Act, the regulations and other laws. HSE inspectors have the power to visit a workplace at any reasonable time and normally want to talk to management, employees, and health and safety representatives. They can look around the premises and inspect documents. If they find something unsafe or against the law, they can

- warn or advise the employer what must be done to put things right
- issue an improvement or prohibition notice. An improvement notice states what the problem is, how to put it right and gives a time limit for complying. A prohibition notice requires the employer to stop doing something until things are put right, and is only issued if there is a risk of serious injury
- prosecute the employer (in Scotland the procurator fiscal decides whether there should be a prosecution) for breaking the law, including failure to comply with an improvement or prohibition notice.

The HSE information line is 0541 545500.

The HSE does not cover all workplaces. It covers factories, building sites, mines, farms, fairgrounds, quarries, railways, chemical plant, offshore and nuclear installations, schools and hospitals. Other workplaces such as retailing, some warehouses, most offices, hotels and catering, sports, leisure, consumer services and places of worship are covered by *local authority enforcement officers* who have similar powers to HSE inspectors. To get in touch with them, contact your local authority environmental health department.

Appendix D

Enforcing the Laws on Workplace Protection

What can you do if your employer refuses to obey the health and safety laws? There are a number of different options, but the lack of effective enforcement of the laws is their biggest weakness.

1. Contact the Health and Safety Executive (see Appendix C) (for factories) or the local authority environment department (for offices and warehouses) and ask them to put pressure on your employer. This could be by advice, or by issuing an improvement or prohibition notice. The HSE or local authority is responsible for bringing a criminal prosecution against employers who do not comply with their legal duty to protect the health and safety of their employees, either by breach of the H&SAWA general duties, or breach of the health and safety regulations, or failure to comply with an improvement notice or prohibition notice. In Scotland the consent of the Procurator Fiscal is needed.

   Unfortunately the HSE inspectors and local authority enforcement officers are very overworked: there are not nearly enough of them to enforce all the health and safety laws in all workplaces.
2. Take out a High Court injunction against an employer restraining him/her from illegally exposing you to a danger. (This would be very expensive and probably impossible without union funding or legal aid, but if successful the dangerous process would be stopped until made safe.)

3. If you or your baby has suffered harm as a result of your employer's failure to obey the health and safety laws, you can sue the employer for damages in the county court. If your baby has been harmed you would have a case under the Congenital Disabilities (Civil Liability) Act 1976. Employers are required by the Employer's Liability (Compulsory Insurance) Act 1969 to hold an Employer's Liability Insurance policy, so that if you win your case, the compensation will be paid regardless of your employer's financial state. Compensation for industrial injuries and diseases is also provided by the State under the Social Security Contributions and Benefits Act 1992. If you are injured at work, or contract one of the diseases that are recognised as being industrial, you have a right to a range of benefits. Negligence by your employer does not have to be proved, but you do have to prove that the injury happened at and arose out of your work.

   No one should have to wait until they suffer harm, but the threat of a claim at some future date might encourage the employer to take the regulations more seriously.
4. If your employer refuses to follow the six steps under the Management of Health and Safety Regulations (see Appendix C), and you decide not to run the risk of carrying on and you stop going to work, you could take a case to an industrial tribunal and argue that your employer has either refused you alternative work or suspended you on maternity grounds but not paid you (see points 5 and 6).
5. If your employer refuses to offer you suitable alternative work, you can apply to the industrial tribunal (within three months) under the Employment Rights Act 1996 section 70(4) for compensation.
6. If your employer suspends you from work but refuses to pay you, you can apply to the industrial tribunal (within three months) for loss of pay under the Employment Rights Act 1996 section 70(1).
7. If your employer says you have to take sick leave because of a health and safety risk at work, you could challenge this by making a claim to an industrial tribunal a) for loss of wages under the Employment Rights Act 1996 section 13, and b) arguing that you have in fact been suspended so your employer should pay you (see 6 above). In the case of *Hickey v Lucas Service UK Ltd*, an industrial tribunal awarded full loss of pay to a woman who went off sick when her employers failed to carry out a risk assessment on her job. The

tribunal held that she was suspended from work even though the doctor had signed her off sick to protect her from possible risks (case reference: IT, Bristol, Case no. 1400979/96).

8. If your employer sacks you because you are unable to do the same work as before your pregnancy, you can make a claim to an industrial tribunal for automatically unfair dismissal (under the Employment Rights Act 1996 section 99(1)) and sex discrimination.
9. If your employer sacks or victimises you for raising safety concerns when there is no safety representative or committee (or there is but it is not reasonably practicable to raise concerns through them), you would have an industrial tribunal claim under the Employment Rights Act 1996 section 44 (victimisation) or section 100 (sacking).
10. If you are a safety representative or representative of employee safety and you are sacked or victimised for carrying out your duties, you would have an industrial tribunal claim under the Employment Rights Act 1996 section 44 (victimisation) or section 100 (sacking).
11. If you leave the workplace because you reasonably believe yourself to be in 'serious and imminent' danger, and your employer sacks or victimises you for this, you would have an industrial tribunal claim under Employment Rights Act 1996 sections 44 and 100.

Update: A 1994 Review of the Health and Safety Regulations by the Health and Safety Commission recommended that 40 per cent of the current regulations cited in the UK Health and Safety at Work Act 1974 be removed in order to update and simplify the UK's health and safety legislation. There is some fear that deregulation and diminution of the role of unions will lead to a lowering of health and safety standards and worker protection.

Appendix E

The Influence of the European Union on Workplace Protection

The European Union (EU) has had a dramatic effect on workplace protection for all workers, but has also paid particular attention to pregnant workers.

THE EUROPEAN UNION WORK RELATED DIRECTIVES

What is a Directive?

A directive is a European law. Once it has been agreed by the member countries of the EU, each country is responsible for passing their own laws to implement the directive.

How are Directives Passed?

A directive is proposed by the European Commission (which is made up of representatives from each country). The Social Questions Working Group, which is composed of representatives from the employment departments of the member states, meet weekly to negotiate an agreement if possible. The proposed directive is then voted on by the Council of Ministers for Social Affairs (this is made up of the employment ministers for the member states).

The directives on health and safety need only what is called a qualified majority vote, or 73 per cent support in the Council of Ministers for Social Affairs. Others require a unanimous vote. Sometimes there is almost as much disagreement on whether an issue is considered to be a health and safety issue or strictly an employment issue as there is over the merits of the proposal itself. The maximum 48-hour working week is one example of such a political 'hot potato'.

Once you understand the process, you can follow the journey of a directive that you are particularly interested in and can make your views known to the governmental agency in charge. Your position will be stronger if you are part of a union or women's organisation.

How Does a Directive Become Law in the UK?

Once a directive is adopted by the European Council of Ministers for Social Affairs, individual countries are given specific dates by which the directive should be implemented. In Britain the Health and Safety Commission studies the directive and publishes consultative documents presenting proposals for its implementation. Interested groups can respond to these proposals which can be amended. The final proposals are drawn up as regulations. The regulations are laid before Parliament for a vote and if accepted they become law. The health and safety directives have been implemented as regulations using powers under the Health and Safety at Work Act 1974 (see Appendix C).

WHAT THE DIRECTIVES OFFER YOU

If you work in the private sector, you have to rely on the UK laws (acts and regulations) for your protection. But if you work in the public sector, you can rely both on the UK laws and directly on the directives themselves. Usually there is not much difference, but this can be useful either before the government has brought in legislation to implement the directive or if the UK legislation has not implemented the directive properly. If the government has not implemented a directive properly, in some circumstances it is possible to sue the government (instead of your employer) if you suffer harm because of this failure (this applies to both private and public sector and is called a 'Francovich' claim).

Here are the highlights of the directives:

The Pregnant Workers Directive 1992

This is the most important piece of European legislation affecting the pregnant worker, covering not only health and safety but also maternity leave and pay:

Health and Safety

Employers must assess their workplaces and if hazards to pregnancy or breastfeeding are found they must inform employees and decide what measures should be taken, i.e.

- temporary adjustment to working conditions and/or hours
- if not feasible, move employee to another job
- if not feasible, employee granted leave for the period needed to protect her and/or foetus
- all contractual employment rights to be maintained during period of leave
- pregnant worker/new mother not to be obliged to perform night work where covered by a medical certificate.

(See Annexes I and II in box below for a non-exhaustive list of potential hazards.)

Maternity Leave

All pregnant women are to be given at least 14 weeks maternity leave irrespective of hours or length of service. This must include compulsory leave of at least two weeks before and/or after confinement. All contractual rights continue during leave.

Dismissal

All pregnant workers are protected against dismissal from the beginning of pregnancy to the end of maternity leave except for exceptional cases not connected to pregnancy or maternity. The employer must give written reasons for dismissal under these circumstances.

Antenatal Care

All pregnant workers are entitled to paid time off to attend antenatal clinics during working hours.

Maternity Pay

During maternity leave workers must be given an 'adequate allowance' to be at least equivalent to sick pay.

This directive has been implemented in the UK as part of the Employment Rights Act 1996, as an amendment to The Management of Health and Safety at Work Regulations 1992 (see Appendix C), and as Statutory Maternity Pay Regulations. See Appendix F for the current UK maternity rights. The directive and its implementation are being reviewed in 1997.

Pregnant Workers Directive Annex I: Non-exhaustive list of Agents, Processes and Working Conditions

These are what the EU identifies as some of the potential hazards for women who are pregnant or have recently given birth or are breastfeeding. They should be included in a risk assessment and trigger the health and safety rights described above.

A. Agents

1. Physical agents where these are regarded as agents causing foetal lesions and/or likely to disrupt placental attachment, and in particular:
   a) shocks, vibration or movement
   b) handling of loads entailing risks, particularly of a dorsolumbar nature
   c) noise
   d) ionising radiation
   e) non-ionising radiation
   f) extremes of cold or heat
   g) movements and postures, travelling – either inside or outside the establishment – mental and physical fatigue and other physical burdens connected with the activity of the worker.
2. Biological agents: Biological agents of risk groups 2, 3 and 4 within the meaning of Article 2(d) numbers 2, 3 and 4 of Directive 90/679/EEC, in so far as it is known that these agents or the therapeutic measures necessitated by such agents endanger the health of pregnant women and the unborn child and in so far as they do not yet appear in Annex II.
3. Chemical agents: The following chemical agents in so far as it is known that they endanger the health of pregnant women and

the unborn child and in so far as they do not yet appear in Annex II:
a) substances labelled R40, R45, R46 and R47 under Directive 67/548/EEC
b) chemical agents in Annex 1 to Directive 90/394/EEC
c) mercury and mercury derivatives
d) amniotic drugs
e) carbon monoxide
f) chemical agents or known and dangerous percutaneous absorption.

Pregnant Workers Directive Annex II

Exposure is prohibited to the following:

A. Pregnant workers

1. Agents
   a) Physical agents
   Work in hyperbaric atmosphere, e.g. pressurised enclosures and underwater diving.
   b) Biological agents
   The following biological agents: toxoplasma, rubella – unless the pregnant workers are proved to be adequately protected against such agents by immunisation.
   c) Chemical agents
   Lead and lead derivatives in so far as these agents are capable of being absorbed by the human organism.
2. Working conditions
   Underground mining work.

B. Workers who are breastfeeding

1. Agents
   a) Lead and lead derivatives in so far as these agents are capable of being absorbed by the human organism.
2. Working conditions
   Underground mining work.

The Framework Directive on Health and Safety 1989

This directive sets out the general goals and duties of workplace protection. It applies to all sectors of work activity. It assigns primary responsibility for health and safety of employees to their employer and sets out general principles for employers to follow in protecting health and safety.

The Health and Safety Commission took the view that as some of the requirements of this directive were already met in the UK by the Health and Safety at Work Act 1974 (see Appendix C), it was better not to change the existing structure. Instead the necessary bits and pieces were added on in the Management of Health and Safety Regulations 1992 (see Appendix C) which were made using powers under the 1974 Act.

There are a number of 'daughter' directives made under the Framework Directive and aimed at specific areas:

The Workplace Directive 1989

This directive applies to most fixed, permanent workplaces and sets the minimum safety and health requirements for the workplace. This includes such things as emergency routes and exits, fire precautions, ventilation, temperature, lighting, dangerous areas, rest rooms, washing facilities, toilets and first aid equipment. The directive states that 'Pregnant women and nursing mothers must be able to lie down to rest in appropriate conditions'.

This directive has been implemented as The Workplace (Health, Safety and Welfare) Regulations 1992 (see Appendix C).

The Use of Work Equipment Directive 1989

This directive applies to all types of work sites and places on the employer the primary responsibility for the health and safety of workers regarding the provision and use of all work equipment. Work equipment is defined as any machine, apparatus, tool or installation used at work.

This directive has been implemented as The Provision and Use of Work Equipment Regulations 1992 (see Appendix C).

The Use of Personal Protective Equipment (PPE) Directive 1989

This directive sets the minimum health and safety requirements for the use by workers of personal protective equipment at the workplace. Personal protective equipment (PPE) should be used only when the risks cannot be avoided or limited by other workplace measures. PPE refers to equipment designed to be worn or held by workers to protect them against one or more hazards likely to endanger their health or safety in the workplace. The employer must provide the PPE which is suitable for the work and fits the worker correctly free of charge, maintain the PPE in clean, good working order, provide information and training in the use of the PPE and involve worker representatives in the selection of the PPE.

This directive has been implemented as The Personal Protective Equipment at Work Regulations 1992 (see Appendix C).

The Manual Handling of Loads Directive 1990

This directive sets the minimum safety and health requirements for the manual handling of loads where there is a risk particularly of back injury to workers. It applies to all industries and services. It advises employers to avoid the need for manual handling or, where manual handling cannot be avoided, to take appropriate steps to reduce or avoid risk of injury. Some of the factors to be taken into account are the size, shape and weight of the load, the physical effort required, the characteristics of the workplace and the frequency of the task. Employees are to provide information training to workers and to consult their representatives when setting up the new procedures.

This directive has been implemented as The Manual Handling Operations Regulations 1992 (see Appendix C).

The Display Screen Equipment (VDUs) Directive 1990

This directive sets the minimum safety and health requirement for work with display screen equipment. Employers installing new workstations need to meet the minimum requirements set out in an annex to the directive. These refer to the design of the display screen, keyboard, desk chair and environmental factors such as glare, lighting, noise, humidity, task design and software. Daily work schedules are to be designed so that work on a display screen is periodically interrupted by breaks or changes of activity.

Workers are entitled to an appropriate eye and eyesight test before starting display screen work and assessment at regular intervals. If they experience visual difficulties attributable to the work, they are entitled to an ophthalmological examination and to special visual corrective appliances if necessary.

This directive has been implemented as The Health and Safety (Display Screen Equipment) Regulations 1992 (see Appendix C).

The Carcinogens Directive

This directive pertains to the protection of workers from the risks related to exposure to carcinogens at work. It applies to those substances labelled 'may cause cancer' and to those substances and processes specified in Annex 1 to the directive. The employer must replace the carcinogen by a less dangerous substitute if at all possible. Failing this, the carcinogen should be used in a closed system. If this is not technically feasible, the exposure of workers must be reduced to as low a level as possible. Monitoring and health surveillance procedures are to be undertaken. The employer is to provide adequate information and training about the risks to health, the precautions to be taken and the results of health surveillance to be provided by the employers.

This directive has been implemented as part of the 1994 COSHH legislation (see Appendix C).

The Biological Agents Directive

The directive pertains to the protection of workers from the risks related to exposure to biological agents at work. It is one of three directives in the general area of microbiology and biotechnology. The directive applies to all work activities in which workers may be exposed to biological agents but makes distinctions between activities where the work is directly involved e.g. microbiological laboratory work, and those where exposure may be incidental to work activity, e.g. health care, farming. Employers have the responsibility to prevent or reduce risk of exposure, to keep lists of exposed workers, and to conduct health surveillance and to provide notification to appropriate authorities.

This directive has been implemented as part of the 1994 COSSH legislation (see Appendix C).

The Asbestos Worker Protection Directive (Amendment)

This directive amends the previous council directive and sets out the framework for the protection of workers from the risks related to exposure to asbestos at work. It applies to all activities in which workers may be exposed in the course of their work to dust arising from asbestos or materials containing asbestos. Implementation in the UK has required only minor amendment to current regulations.

The Temporary Workers Directive 1991

This directive supplements the measures to encourage improvements in the safety and health at work of temporary employees including agency workers and fixed-duration employees in the framework directive.

This directive has not been fully implemented. Most of the MHSW Regulations apply to temporary workers, but of the 'six steps' for protecting pregnant workers and new mothers, steps 4–6 only apply to 'employees' i.e. not to agency workers. The PPE at Work Regulations and the Manual Handling Operations Regulations also only apply to 'employees', so the duties would fall on the employment agency not the business where they are working – in breach of this directive.

The European Health and Safety Agency

The purpose of this agency, located in Bilbao, Spain, is to promote improvements in worker health and safety in the European Union countries. It aims to provide technical, scientific and economic information needed for upgrading health and safety in the workplace. This includes information on research, preventative activities and available training programmes. The European Health and Safety Agency sponsors an information network. In the UK, the Health and Safety Executive receives and distributes information from the Agency and to and from relevant British health and safety organisations.

This information network will also identify centres that have particular expertise in specific health and safety topics.

The European Year of Safety, Hygiene and Health in the Workplace

The Council of Ministers of the European Union declared 1992 the European Year of Safety, Hygiene and Health in the Workplace. It ran from 1 March 1992 and ended on 28 February 1993. The main aim was to raise awareness among workers and employers about health and safety problems in the workplace and the best way to deal with them. Another important feature was to stress the economic as well as the social importance of maintaining high standards of health and safety at work. A national committee was set up by the Health and Safety Commission (HSC) to coordinate the UK's response to the European initiative. It sought to achieve maximum publicity for occupational health and safety related activities especially for the four themes the EU identified for particular emphasis: clean air at work; safe working practices; well-being at work and noise and vibration. These four aspects of the workplace are important for all workers, but are especially important for the pregnant worker.

Appendix F

Other Maternity Rights

This appendix sets out statutory maternity rights other than the health and safety rights. Most of these are contained in the Employment Rights Act 1996, others are in the maternity pay regulations. These rights are the legal minimum but you should always check with your employer, personnel department or trade union representative to see whether better rights have been agreed at your workplace.

These rights are accurate as at October 1997. However, the law in this area changes frequently so for up-to-date information send a stamped, self-addressed envelope enclosing a cheque or postal order for £1.00 to Maternity Alliance, 45 Beech Street, London EC2P 2LX, and ask for the leaflet *Pregnant at Work*.

TIME OFF FOR ANTENATAL CARE

- All pregnant women have the right not to be unreasonably refused any time off work they need for antenatal care and to be paid for such time off. This means that you can take time off for your antenatal appointments, including time needed to travel to your clinic or GP, without loss of pay. All pregnant women have this right, whatever hours they work and however recently they started their job.
- You should let your employer know when you need time off and how long you are likely to be away. For appointments after the first one, your employer can ask to see your appointment card and a certificate signed by your GP, midwife or health visitor, stating that you are pregnant.

- Antenatal care includes parentcraft and relaxation classes so you should get paid for any time off you need to go to these classes. You may need a letter to show your employer from your GP, midwife or health visitor, saying these classes are part of your antenatal care.
- If you are not paid or are refused time off you can make a claim to the industrial tribunal within three months.

MATERNITY LEAVE

Fourteen weeks maternity leave

Who gets it?

- Every woman who is employed while she is pregnant is entitled to 14 weeks maternity leave, provided she gives her employer proper notice (see below). It doesn't matter how many hours you work and it doesn't matter how long you have worked for your employer.

How do I get it?

- Write to your employers at least 21 days before you start your maternity leave telling them a) that you are pregnant and b) the expected week of childbirth. You must also tell them c) the date on which you intend to start your maternity leave (you only need to put this in writing if your employers ask for it). Finally you must, if your employers ask for it, d) enclose a copy of your maternity certificate (form MAT B1) which your midwife or GP will give you when you are about six months pregnant. If you cannot give 21 days notice, for example, if you have to go into hospital unexpectedly, you must write to your employers as soon as you reasonably can.

When can I start it?

- The earliest you can start your maternity leave is eleven weeks before the expected week of childbirth. It is for you to decide when you want to stop work. You can even work right up until the day of childbirth. The only exception to this rule is that if you have a pregnancy-related illness/absence in the last six weeks of your pregnancy, you may be bounced onto your maternity leave even if you are absent for only one day. However if you are ill only for a short time your employer may agree to let you start your maternity leave when you had planned. If your baby is born early, before the day you were planning to start your leave, then leave will start on the day of birth.

What will I get while I'm away?

- During the 14-week maternity leave period all your contractual rights (including holiday entitlement and benefits such as a company car) will continue as if you were not absent from work, except your normal pay. See below for details of maternity pay.

What happens when I go back?

- You do not need to give any notice of return if you are going back to work at the end of your 14-week maternity leave period (14 weeks from the day you began your leave). When you go back it will be to exactly the same job.
- If you want to return to work early you must give your employers seven days' notice in writing of the date you will be returning. If you do not give this notice and just turn up at work before the end of the 14-week period, your employers can send you away for seven days or until the end of your 14-week period, whichever is earlier.
- The law does not allow you to work for two weeks after childbirth so if your baby is born very late and your 14-week maternity leave period has run out, your maternity leave is extended by two weeks from the actual birth of your baby.
- If you are unable to go back to work at the end of your 14-week maternity leave period because of sickness, you can remain off work for up to four weeks longer and still be protected from being unfairly dismissed if you have given your employer a medical certificate before the end of your maternity leave. If you are receiving Statutory Maternity Pay (see below) you will continue to receive it for these four weeks. If your employer has a contractual sick pay scheme you should get sick pay for these weeks.

Extended maternity absence

(The right to return to work up to 29 weeks from the week of childbirth)

Who gets it?

- Women who have worked for the same employer for at least two years by the end of the twelfth week before the week the baby is due. If you work in a firm which employs five or fewer people you do not have a clear right to return after 29 weeks even if you have the service, because your employer does not have to allow you to return to work if it is 'not reasonably practicable' a) for you to have your job back *or* b) to give you suitable alternative work on terms and conditions which are not substantially less favourable to you. However, you still qualify for 14 weeks maternity leave (see above).

How do I get it?

- As well as giving the notice for your 14 weeks maternity leave, see a–d above, you must also e) tell your employers that you intend to take extended maternity absence and return to work after the birth. (This does not mean you have to return – you can change your mind later.) You must put this in writing at least 21 days before you start your leave, but you do not have to give a date for your return.

When can I start it and how long does it last?

- Extended maternity absence is an 'add on' to maternity leave. The first 14 weeks you are away will be ordinary maternity leave, so the same rules apply about choosing when to leave. The earliest you can leave work is the beginning of the eleventh week before the *expected* week of childbirth. You have to return to work *by the end of the 29th week from the actual week of childbirth.* (Remember to count 29 weeks from the Sunday at the beginning of the week in which your baby is born.)

What will I get while I'm away?

- The first 14 weeks that you are away will be ordinary maternity leave and so the same rules apply about your rights (see above).

The rest of the time you are off is extended maternity absence and you should ask for your contractual rights to continue.

What happens when I go back?

- Your employers may write to you any time from eleven weeks after the start of your maternity leave asking you to confirm that you are going back to work. You must reply in writing within 14 days clearly saying that you are going back. This does not mean you have to go back: you can still change your mind later. It simply keeps open your option to return, otherwise you will lose that right. You don't have to give a date of return at this stage. If your employer doesn't write to you, you don't have to do anything.
- At least 21 days before you intend to return you must write to your employers giving notice of the exact date of your return. This is called your Notified Date of Return.
- When you return to work after extended maternity absence your employer must give you back the same job or, if that is not reasonably practicable, a suitable job on similar terms and conditions.
- You cannot delay your return beyond the Notified Date of Return except in the following four situations:
- If you are ill you can delay going back for up to four weeks. You must let your employers know before the date on which you intended to return that you will be extending your leave because of sickness, and you must send in a medical certificate.
- Your employers can also delay your return for up to four weeks. They must tell you the reason for the delay and give you a new date for your return.
- An interruption of work, such as a strike, which stops you returning to work allows you to delay your return until work starts again. If the interruption stops you giving notice of the date you intend to return, you can delay your return for up to 28 days after the end of the interruption.
- You may delay your return by agreement with your employer.

MATERNITY PAY AND ALLOWANCE

Statutory Maternity Pay (SMP)

What is it?

- A weekly payment for women employed during pregnancy. Your employers pay it to you and then they claim most of it back from the Inland Revenue because it is really a state benefit. You can get it even if you don't plan to go back to work. You will not have to pay it back if you don't return to work.

Who gets it?
Women who:

- have worked for the same employer for at least 26 weeks by the end of the qualifying week (the 15th week before the week the baby is due). To find out which is the qualifying week, look on a calendar for the Sunday before your baby is due (or the day it is due if that is a Sunday) and count back 15 Sundays from there. You must have worked at least 26 weeks with your employers by the end of that week in order to qualify for SMP.
- are still in their job in this qualifying week. (It doesn't matter if you are off work sick, or on holiday.)
- earn £62 per week or more on average in the eight weeks (if you are weekly paid) or two months (if you are monthly paid) before the end of the qualifying week. This is the 1997 rate.

If you are not sure if you're entitled to SMP, ask anyway. Your employer will work out whether or not you should get it.

How much is it?

- For the first six weeks you get 90 per cent of your average pay. (The average is calculated from the pay you actually received in the eight weeks or two months before the last pay day before the end of the qualifying week.) After that you get the basic rate of SMP which is currently (1997) £55.70 per week for up to twelve weeks.

- You won't get SMP for any week that you work. So, if you qualify for only 14 weeks maternity leave and you return to work at the end of the 14th week you will lose the last 4 weeks of your 18 weeks of SMP.

When is it paid?

- SMP is paid for up to 18 weeks and week 11 before the expected week of childbirth is the earliest it can start. It always starts on the Sunday after you go on maternity leave. So if your last day of work is a Friday or Saturday it will start immediately.
- It is for you to decide when you want to stop work. You can work right up until the week of childbirth and you will not lose SMP. The only exception to this rule is that if you have a pregnancy-related illness in the last six weeks of your pregnancy, you will not be entitled to claim Statutory Sick Pay and will start receiving your SMP. If your illness is not pregnancy-related you can claim Statutory Sick Pay in the last six weeks and start your SMP when you had planned to.
- Your employers should pay your SMP. They deduct any tax and National Insurance contributions.

How do I get it?

- Write to your employers at least 21 days before you plan to stop work, telling them the date you are going on leave and asking for SMP. You must also send them a copy of the maternity certificate (form MAT B1) which your GP or midwife will give you when you are about six months pregnant.

Maternity Allowance

What is it?

- A weekly allowance for women who work just before or during their pregnancy but who can't get SMP. You can get Maternity Allowance if you are self-employed or if you gave up work or changed jobs during your pregnancy.

Who gets it?

- Women who can't get SMP but who have worked and paid full rate National Insurance contributions for at least 26 of the 66 weeks before the expected week of childbirth. If in doubt, claim. Your local benefits agency will work out whether or not you can get the benefit.

How much is it?

- There are two rates. If you are self-employed or unemployed in the qualifying week (the 15th week before the expected week of childbirth), you get £48.35 per week. If you are employed in the qualifying week you get £55.70 per week. There are no deductions for tax or National Insurance. These are 1997 rates.

When is it paid?

- Maternity Allowance is paid for up to 18 weeks. It is only paid for weeks when you are not working. The earliest it can start is 11 weeks before the expected week of childbirth. If you are employed or self-employed you can choose when to start your Maternity Allowance in the same way as for SMP. However if you are unemployed your Maternity Allowance must start in week 11 before the expected week of childbirth.

How do I claim?

- Fill in form MA1 (available from the Benefits Agency or your antenatal clinic) and send it to your local Benefits Agency along with your maternity certificate (form MAT B1) from your GP or midwife, and (if you are employed) form SMP1 which your employer will give you if you do not qualify for SMP. Send in form MA1 as soon as you can after you are 26 weeks pregnant. Don't put off sending it in because you are waiting for the MAT B1. You can send it later.
- If you have not paid 26 weeks' National Insurance contributions by the time you are 26 weeks pregnant, then you may decide to work later into your pregnancy and you should send off the

MA1 form as soon as you have made 26 weeks' National Insurance contributions.

DISMISSAL OR UNFAIR TREATMENT

- It is against the law for your employer to dismiss you or select you for redundancy for any reason connected with pregnancy, childbirth or maternity leave. It does not matter how long you have worked for your employer or how many hours a week you work.
- If you are dismissed while you are pregnant or during your maternity leave, your employer must give you a written statement of the reasons.
- If you are dismissed and you believe the reason is connected with your pregnancy, get advice from your trade union or Citizen's Advice Bureau. You must put in your claim to the industrial tribunal within three months of dismissal.
- If you are sacked or are treated unfairly (e.g. having your pay cut) because you are pregnant or have had a baby, you may have a claim for sex discrimination. It doesn't matter how long you have worked for your employer. The law is complicated, so get advice. These claims also have to be made within three months.

Useful Addresses

ORGANISATIONS AND INFORMATION RESOURCES

Bradford Occupational Health Project, 23 Harrogate Road, Bradford 2. Tel: 01274 626191.

British RSI Association, Chapel House, 152–156 High Street, Yiewsley, West Drayton, Middlesex UB7 7BE. Tel: 01895 431134.

Business in the Community, 8 Stratton Street, London W1X 5FD. Tel: 0171 629 1600.

Camden Occupational Health Project, Bloomsbury Health Education Department, St Pancras Way, London NW1. Tel: 0171 383 0997.

Campaign for Freedom of Information, 88 Old Street, London EC1V 9AR. Tel: 0171 253 2445.

National Association of Citizens Advice Bureaux, 115 Pentonville Road, London N1. Tel: 0171 833 2181.

Commission of the European Communities, 8 Storey's Gate, London SW1P 3AT. Tel: 0171 973 1992.

Confederation of British Industry, Centre Point, 103 New Oxford Street, London WC1A 1DU. Tel: 0171 379 7400.

Daycare Trust/National Childcare Campaign, 4 Wild Court, London WC2B 5AU. Tel: 0171 405 5617.

Department of Employment Library (Reference and Enquiries Librarian), Steel House, Tothill Street, London SW1H 9NF. Tel: 0171 273 4711.

Education & Employment Law Information Service, Hamilton House, Mabledon Place, London WC1. Tel: 0171 383 2585.

Environmental Health Officers (local authorities), Equal Opportunities Commission, Library and Information Services, Overseas House, Quay Street, Manchester M3 3HN. Tel: 0161 833 9244.

Equal Opportunities Commission (Press office), 36 Broadway, London SW1H 0XH. Tel: 0171 222 1110.

Eyecare Information Service, PO Box 3597, London SE1 6DY. Tel: 0171 357 7730.

Greater Manchester Hazards Centre, 23 New Mount Street, Manchester M4 4DE. Tel: 0161 953 4037.

Health Education Authority, Hamilton House, Mabledon Place, London WC1H 9TX. Tel: 0171 383 3833.

Health and Safety Advice Centre, Unit 304, The Argent Centre, 60 Frederick Street, Birmingham, B1 3HS. Tel: 0121 236 0801.

Health and Safety Project, Trade Union Studies Information Unit, Southend, Fernwood Road, Jesmond, Newcastle NE2 1TJ. Tel: 0191 281 4217.

HMSO Publications Centre, 51 Nine Elms Lane, London SW8 5DR. Tel: 0171 873 0011.

HMSO Bookshop (counter service only), 49 High Holborn, London WC1V 6HB. Tel: 0171 873 0011.

HMSO Bookshop, 68–69 Bull Street, Birmingham B4 6AP. Tel: 0121 236 9696.

HMSO Bookshop, Southey House, 33 Wine Street, Bristol BS1 2BQ. Tel: 0117 926 4306 .

HMSO Bookshop, 9–21 Princess Street, Manchester M60 8AS. Tel: 0161 834 7201.

HMSO Bookshop, 16 Arthur Street, Belfast BT1 4GD. Tel: 01232 238451.

Institution of Environmental Health Officers (EHO), 15 Hatfields, London SE1. Tel: 0171 928 6006.

Institute of Occupational Health, PO Box 363, Birmingham B15 2TT.

Institute of Occupational Safety and Health (IOSH), The Grange, Highfield Drive, Wigston, Leicester LE18 1NN. Tel: 0116 257 1399.

Issue – The National Fertility Association, 509 Aldridge Road, Great Barr, Birmingham B44 8NA. Tel: 0121 344 4414

Fertility Helpline. Tel: 0121 344 4414 (information given is free of charge).

Kids' Clubs Network, Bellerive House, 3 Muirfield Crescent, London E14 9SZ. Tel: 0171 512 2112.

Labour Research Department (LRD), 78 Blackfriars Road, London SE1 8HF. Tel: 0171 928 3649.

Liverpool Occupational Health Project, 18–20 Manchester Street, Liverpool L1 6ER.

London Hazards Centre, Headland House, 308 Grays Inn Road, London WC1X 8DS. Tel: 0171 837 5605.

Maternity Alliance, 45 Beech Street, London EC2P 2LX. Tel: 0171 588 8583.

National Childbirth Trust, Alexander House, Oldham Terrace, London W3 6NH. Tel: 0181 992 8637.

National Childminding Association, 8 Masons Hill, Bromley, Kent BR2 9EY. Tel: 0181 464 6164.

National Council for One Parent Families, 255 Kentish Town Road, London NW5 2LX. Tel: 0171 267 1361.

New Ways to Work, 309 Upper Street, London N1 2TY. Tel: 0171 226 4026.

Parents at Work, 45 Beech Street, London EC2P 2LX. Tel: 0171 628 3578.

Portsmouth Area Health and Safety Group, 32 Rowner Close, Gosport, Hants PO13N 0LY.

Royal Society for Prevention of Accidents, Edgbaston Park, 353 Bristol Road, Birmingham B5 7ST. Tel: 0121 248 2000.

Society of Occupational Medicine, Royal College of Physicians, 6 St Andrews Place, Regents Park, London NW1 4LE. Tel: 0171 487 3414.

Society of Occupational Health Nurses, Royal College of Nursing, 20 Cavendish Square, London W1M 9AE. Tel: 0171 409 3333.

VDU Workers' Rights Campaign, c/o City Centre, 32–35 Featherstone St., London EC1Y 8QX. Tel: 0171 608 1338 (contact for publications, health and safety, training etc.).

Walsall Action for Safety and Health, 7 Edinburgh Drive, Rushall, Walsall WS4 1HW. Tel: 01922 25860.

Women's Design Service, 4 Pinchin Street, London E1. Tel: 0171 709 7910 (contact for information and resources on women and the built environment).

Women's Health and Reproductive Rights Information Centre, 52 Featherstone Street, London EC1Y 8RT. Tel: 0171 251 6580.

Working for Childcare Consultancy Service, 77 Holloway Road, London N7 8JF. Tel: 0171 700 0281.

Electronic Mail Networks

Poptel (Geonet), 25 Downham Road, London N1. Tel: 0171 249 2948.

GreenNet, 23 Bevenden St., London N1 6BH. Tel: 0171 608 3040.

Libraries

The British Library Science Reference & Information Service (BLSRIS), 25 Southampton Buildings, Chancery Lane, London WC2A 1AW. Tel: 0171 323 7494.

Combes Medical Library, Ealing Hospital, Central Reference Library, 103 Ealing Broadway Centre, London W5 5JY. Tel: 0181 567 3656.

Redbridge Libraries Health and Safety Collection, Central Reference Library, Clements Road, Ilford, Essex IG1 1EA. Tel: 0181 478 7145.

Westminster Health Information Library, Marylebone Road, London NW1. Tel: 0171 798 1039.

HEALTH AND SAFETY EXECUTIVE OFFICES AND THE EMPLOYMENT MEDICAL ADVISORY SERVICE

Health and Safety Executive (HQ), Rose Court, 2 Southwark Bridge, London SE1 9HS. Tel: 0171 717 6000.

The South and East

Health and Safety Executive, East Anglia Area Office, 39 Baddow Road, Chelmsford, Essex CM2 0HL. Tel: 01245 706200.

Health and Safety Executive, London North Area Office, Maritime House, 1 Linton Road, Barking, Essex IG11 8HF. Tel: 0181 235 8000.

Health and Safety Executive, London South Area Office, 1 Long Lane, London SE1 4PG. Tel: 0171 556 2100.

Health and Safety Executive, Northern Home Counties Area Office, 14 Cardiff Road, Luton, Bedfordshire LU1 1PP. Tel: 01582 444200.

Health and Safety Executive, South Area Office, Priestley House, Priestley Road, Basingstoke RG24 9NW. Tel: 01256 404000.

Health and Safety Executive, South East Area Office, East Grinstead House, East Grinstead, West Sussex RH19 1RR. Tel: 01342 334200.

Health and Safety Executive, South West Area Office, Inter City House, Mitchell Lane, Victoria Street, Bristol BS1 6AN. Tel: 0117 988 6000.

The Midlands and Wales

Health and Safety Executive, East Midlands Area Office, Belgrave House, Greyfriars, Northampton NN1 2BS. Tel: 01604 738300.

Health and Safety Executive, The Marches Area Office, The Marches House, Midway, Newcastle under Lyme, Staffordshire ST5 1DT. Tel: 01782 602400.

Health and Safety Executive, North Midlands Area Office, The Pearson Building, 55 Upper Parliament Street, Nottingham NG1 6AU. Tel: 01159 712800.

Health and Safety Executive, Wales Area Office, Brunel House, 2 Fitzalen Road, Cardiff CF2 1SH. Tel: 01222 263000.

Health and Safety Executive, West Midlands Area Office, McLaren Building, 35 Dale End, Birmingham B4 7N. Tel: 0121 607 620.

The North and Scotland

Health and Safety Executive, Greater Manchester Area Office, Quay House, Quay Street, Manchester M3 3JB. Tel: 0161 952 8200.

Health and Safety Executive, Merseyside Area Office, The Triad, Stanley Road, Bootle L20 3PG. Tel: 0151 479 2200 .

Health and Safety Executive, North East Regional Office, Arden House, Regent Centre, Gosforth, Newcastle upon Tyne NE3 3JN. Tel: 0191 202 6200.

Health and Safety Executive, North West Area Office, Victoria House, Ormskirk Road, Preston PR1 1HH. Tel: 01772 836200.

Health and Safety Executive, Scotland East Area Office, Belford House, Belford Road, Edinburgh EH4 3UE. Tel: 0131 247 2000.

Health and Safety Executive, Scotland West Area Office, 375 West George Street, Glasgow G2 4LW. Tel: 0141 275 3000.

Health and Safety Executive, South Yorkshire Area Office, Sovereign House, 110 Queen Street, Sheffield S1 2ES. Tel: 0114 291 2300.

Health and Safety Executive, West and North Yorkshire Area Office, 8 St Paul's Street, Leeds LS1 2LE. Tel: 0113 283 4200.

Other HSE Services

Employment Medical Advisory Service. Tel: 0171 221 0870 (contact for detailed information).

HSE Library and Information Services, Public Enquiry Point, Broad Lane, Sheffield S3 7HQ. Tel: 01142 752539.

HSE Books, PO Box 1999, Sudbury, Suffolk CO10 6FS. Tel: 01787 881165.

TRADE UNIONS

Amalgamated Engineering Union (AEU), Hayes Court, West Common Road, Bromley, Kent BR2 7AU. Tel: 0181 462 7755.

Banking, Insurance and Finance Union (BIFU), Sheffield House, 1b Amity Grove, Raynes Park, London SW20 0LG. Tel: 0181 946 9151.

Birmingham Region Union, Safety and Health Campaign (BRUSH), Unit 304, The Argent Centre, 60 Frederick Street, Birmingham B1 3HS. Tel: 0121 236 0801.

Civil and Public Services Association (CPSA), 160 Falcon Road, London SE11 2LN. Tel: 0171 924 2727.

GMB, 22–24 Worple Road, London SW19 4DD. Tel: 0181 947 3131.

Manufacturing, Science and Finance Union (MSF), Park House, 64–66 Wandsworth Common North Side, London SW18 2SH. Tel: 0181 871 2100.

National Communications Union (NCLU), Greystoke House, 150 Brunswick Road, London W5 1AW. Tel: 0181 998 2981.

National and Local Government Officers' Association (NALGO), 1 Mabledon Place, London WC1H 9AJ. Tel: 0171 388 2366.

National Union of Teachers (NUT), Hamilton House, Mabledon Place, London WC1H 9BD. Tel: 0171 388 6191.

Trades Union Congress (TUC), South East Regional Office, Congress House, Great Russell Street, London WC1B 3LS. Tel: 0171 636 4030 (contact for information on regional councils of the TUC).

Union of Communication Workers (UCW), UCW House, Crescent Lane, Clapham, London SW4 9RN. Tel: 0171 622 9977 .

Union of Construction, Allied Trades and Technicians (UCATT), UCATT House, 177 Abbeville Road, London SW4 9RL. Tel: 0171 622 2442.

Union of Shop, Distributive and Allied Workers (USDAW), 188 Wilmslow Road, Fallowfield, Manchester M14 6JL. Tel: 0161 224 2804.

Union of Textile Workers, Foxlowe, Market Place, Leek, Staffordshire ST13 6AD. Tel: 01538 382068.

UNISON (Union of public sector workers), Civic House, 20 Grand Depot Road, London SE18 6SF. Tel: 0171 388 2366.

References/Suggested Reading

Ahlborg, G. Jr., Bodin, L. and Hogstedt, C. 'Heavy Lifting During Pregnancy – A Hazard to the Fetus? A Prospective Study', *International Journal of Epidemiology*, 1990; 142:1241–4.

Ashford, N. and Miller, C. *Chemical Exposures: Low Levels and High Stakes*. New York: Van Nostrand Reinhold, 1991.

Axelsson, G. and Rylander, R. 'Outcome of Pregnancy in Women Engaged in Laboratory Work at a Petrochemical Plant', *American Journal of Industrial Medicine*, 1989; 16:539–45.

Bamford, C. and McCarthy, C. *Women Mean Business: A Practical Guide for Women Returners*. London: BBC Books, 1991.

Barlow, S. M. and Sullivan, F. M. *Reproductive Hazards of Industrial Chemicals: An Evaluation of Animal and Human Data*. London: Academic Press, 1982.

Bernhardt, J.H. 'Potential Workplace Hazards to Reproductive Health. Information for Primary Prevention', *Journal of Obstetrics, Gynecology and Neonatal Nursing*, 1990; 19:53–62.

Bibby, P., ed. *Personal Safety for Social Workers*. Aldershot: Arena, 1994.

Blackwell, R. and Chang, A. 'Video Display Terminals and Pregnancy. A Review', *British Journal of Obstetrics and Gynaecology*, 1988; 95:446–53.

Bonde, J.P. 'Semen Quality and Sex Hormones Among Mild Steel and Stainless Steel Welders: A Cross Sectional Study', *British Journal of Industrial Medicine*, 1990; 47:508–14.

Bonde, J.P. 'Semen Quality in Welders Before and After Three Weeks of Non-Exposure', *British Journal of Industrial Medicine*, 1990; 47:515–8.

Brannen, J. and Moss, P. *Managing Mothers: Dual Earner Households After Maternity Leave*. London: Unwin Hyman, 1991.

Bregmen, D.T., Anderson, K.E., Buffler, P. and Salg, J. 'Surveillance for Work-Related Adverse Reproductive Outcomes', *American Journal of Public Health*, 1989; 79 Suppl:53–7.

British Department of Employment. Employment Department Group. *The Best of Both Worlds: The Benefit of a Flexible Approach to Working Arrangements*. London: Department of Employment, 1991.

Canadian Pediatric Society, Infectious Diseases and Immunization Committee. 'Cytomegalovirus Infection in Day-Care Centres: Risks to Pregnant Women', *Canadian Medical Association Journal*, 1990; 142:547–9.

Cardy, C. in association with Lamplugh, D. *Training for Personal Safety at Work*. Aldershot: Gower for the Suzy Lamplugh Trust, 1995.

Center for Science in the Public Interest. *The Household Pollutant Guide*. New York: Anchor, 1978.

Central Office of Information. *The Best of Both Worlds: The Benefits of a Flexible Approach to Working Arrangements: A Guide for Employers*. London: Employment Department, 1991.

City Centre. *Racial Harassment at Work: A City Centre Guide*. London: City Centre, 1991.

City Centre. *We're Counting on Equality: Monitoring Equal Opportunities in the Workplace in Relation to Sex, Race, Disability, Sexuality, HIV/AIDS and Age*. London: City Centre, 1995.

City Centre/USDAW. *The 'Six Pack': USDAW Guide to the 1992 Health & Safety Regulations*. London: City Centre, USDAW, 1992.

Clark, N., Cutter, T. and McGrane, J.A. *Ventilation: A Practical Guide*. New York: Center for Occupational Hazards, 1984.

Cohen, B. *Caring for Children, Services and Policies for Childcare and Equal Opportunities in the United Kingdom*. London: Commission of the European Communities, 1988.

Coleman, M. and Beral, V. 'A Review of Epidemiological Studies of the Health Effects of Living Near or Working with Electricity Generation and Transmission Equipment', *International Journal of Epidemiology*, 1988; 17:1–13.

Collins, H. *The EU Pregnancy Directive: A Guide for Human Resource Managers*. Oxford: Blackwell Business, 1994.

Commission of the European Communities. Directorate-General for Employment. *Social Europe: The Labour Market Initiatives and Texts Adopted in the Social Field in 1990*. 3. Commission of the European Communities. Directorate-General for Employment, 1990.

Commission of the European Communities. Directorate-General for Employment. *Social Europe: Health and Safety Protection at the Workplace*. 2. Commission of the European Communities. Directorate-General for Employment, 1990.

Cox, S. *Maternity Rights*. London: Industrial Relations Society (IRS), 1996.

Craig M. *Office Workers' Survival Handbook: A Guide to Fighting Health Hazards in the Office*. London: BSSRS Publications, Ltd., 1981.

Daniell, W.E., Vaughan, T.L. and Millies, B.A. 'Pregnancy Outcomes Among Female Flight Attendants', *Aviation, Space and Environmental Medicine*, 1990; 61:840–4.

Daniels, C., Paul, M. and Rosofsky, R. *Family, Work and Health*. Boston: Women's Health Unit, Dept. of Public Health, 1988.

Daniels, C., Paul, M. and Rosofsky, R. 'Health, Equity, and Reproductive Risks in the Workplace', *Journal of Public Health Policy*. 1990; 4:449–62.

Daniels, L. and Shooter, R. *Employer's Guide to Childcare*. New ed. London: The Working Mothers Association, 1991.

Danse, I.R. *Common Sense Toxics in the Workplace*. New York NY: Von Nostrand Reinhold, 1991.

Department of Employment. *Sexual Harassment in the Workplace: A Guide for Employers*. London: Department of Employment, 1992.

Department of Trade and Industry. *Maternity Rights: A Guide for Employers and Employees*. Department of Trade and Industry. Code PL958. (from DTI DSS).

Eaton, J. and Gill, C. *The Trade Union Directory: A Guide to all TUC Unions*. London: Pluto Press, 1983.

Equal Opportunities Commission. *Equal Treatment for Men and Women: Strengthening the Acts*. Manchester: Equal Opportunities Commission, 1988.

Florack, E. and Zielhuis, G. 'Occupational Ethylene Oxide Exposure and Reproduction', *International Archive of Occupational and Environmental Health*, 1990; 62:273–7.

Fraser, T. Morris, *Ergonomic Principles in the Design of Hand Tools*, Occupational Safety and Health Series, No.44. Geneva: International Labour Office, 1980.

Gingerbread. *Starting a Creche: A Guide for Local Community Groups*. London: Gingerbread, 1987.

Green, J. 'Detecting the Hypersusceptible Worker: Genetics and Politics in Industrial Medicine', *International Journal of Health Services*, 1983; 13:247–64.

Guirguis, S., Pelmear, F., Roy, M. and Wong, L. 'Health Effects Associated with Exposure to Anaesthetic Gases in Ontario Hospital Personnel', *British Journal of Industrial Medicine*, 1990; 47:490–7.

Harris, D., ed. *Noise Control Manual: Guidelines for Problem-Solving in the Industrial/Commercial Acoustical Environment*. Florence, Kentucky: Van Nostrand Reinhold, 1991.

Hart, R.W., Terturro, A. and Nimeth, L., eds. 'Report of the Consensus Workshop on Formaldehyde', *Environmental Health Perspectives*, 1984; 58:323–81.

Health and Safety Executive. *HIV and AIDS: A Guide for Branches*. London: Health and Safety Executive.

Health and Safety Executive. *Homeworking: Guidance for Employers and Employees on Health and Safety*. London: Health and Safety Executive.

Health and Safety Executive. *Managing Health and Safety in Schools*. London: Health and Safety Executive.

Health and Safety Executuve. *New and Expectant Mothers: A Guide for Employers*. London: Health and Safety Executive, updated regularly.

Health and Safety Executive. *Violence to Staff*. London: Health and Safety Executive, 1991.

Health and Safety Executive. *Working Alone in Safety*. London: Health and Safety Executive, 1991.

Health and Safety Executive. *Health and Safety Regulation: A Short Guide*. London: Health and Safety Executive, 1995.

Helmkamp, T.E. *et al.* 'Occupational Noise Exposure, Noise-Inducing Hearing Loss, and the Epidemiology of High Blood Pressure', *American Journal of Epidemiology*, 1985; 121:501–14.

Hogg, C. *Under five and Under Funded*. London: The National Childcare Campaign/Daycare Trust, 1988.

Huel, G., Mergler, D. and Bowler, R. 'Evidence for Adverse Reproductive Outcomes Among Women Microelectronic Assembly Workers', *British Journal of Industrial Medicine*, 1990; 47:400–4.

Huff, J.A. 'Carcinogenicity of Select Organic Solvents', in *Proceedings of the International Conference on Organic Solvent Toxicity*, Stockholm, Oct. 1984.

International Agency for Research on Cancer. *IARC Monographs on the Evaluation of the Carcinogenic Risk of Chemicals to Humans. Chemicals, Industrial Processes and Industries Associated with Cancer in Humans*, Supplement 4, International Agency for Research on Cancer, Lyon, 1982; 7–264.

International Labor Office. *Safety and Health in the Use of Agrochemicals: A Guide*. Albany, New York: International Labor Office, 1991.

Kaye, L. *Reproductive Hazards in the Workplace: Some Case Studies*, National Action Committee on the Status of Women, Toronto, Ontario, Canada, 1986.

Kenen, R. *Reproductive Hazards in the Workplace: Mending Jobs, Managing Pregnancies*. Binghamton New York: Haworth, 1993.

Klitzman, S., Silverstein, B., Punnett, L. and Mock, A. 'A Women's Occupational Health Agenda for the 1990s', *New Solutions*, 1990; 1.

Koren, H. *Illustrated Dictionary of Environmental Health and Occupational Safety*. London: Lewis, 1996.

Kozak, M. *Daycare for Kids: A Parents' Survival Guide*. London: The Daycare Trust, 1989.

Kyyronen, P., Taskinen, H., Lindbohm, M.L., Hemminki, K. and Heinonen, O.F. 'Spontaneous Abortions and Congenital Malformations Among Women Exposed to Tetrachloroethylene in Dry Cleaning', *Journal of Epidemiology and Community Health*, 1989; 43:346–51.

La Rosa, J.H. *Women, Work and Health: Employment as a Risk Factor for Coronary Heart Disease*, 1988; 158:6 Part 2:1597–1602.

Labour Research Department. *1992 & Beyond: A Trade Unionists Guide to Developments in the European Community*. London: Labour Research Department, 1991.

Labour Research Department. *Safety Reps Action Guide*. London: Labour Research Department, 1991.

Labour Research Department *Workplace Health: A Trade Unionists' Guide*. London: Labour Research Department, 1989.

LaDou, J., ed. 'The Microelectronics Industry', *Occupational Medicine: State of the Art Reviews*, 1:1 (entire issue) 1986.

Leighton, P. *The Work Environment: The Law of Health, Safety and Welfare*. London: Industrial Society Press, 1991.

Levy, B.S., 'Recognizing and Preventing Hazards in the Workplace', *Consultant*, Nov. 1983: 63–78.

Levy, B.S. 'The Occupational History, Occupational Health Committee', *Annals of Internal Medicine*, Nov. 1983; 99, 8: 643–4.

Levy, B.S. and Wegman, D.H. eds. *Occupational Health: Recognizing and Preventing Work Related Disease*. Boston, Little Brown, 1995.

Lifton, B. *Bug Busters: Getting Rid of Household Pests Without Chemicals*. New York: Avery Publishing Group, 1991.

Lindbohm, M.L., Taskinen, H., Sallmen, M. and Hemminki, K. 'Spontaneous Abortion Among Women Exposed to Organic Solvents', *American Journal of Industrial Medicine*, 1990; 17:449–63.

London Hazards Centre. *Basic Health and Safety: Workers' Rights and How to Win Them*. London: London Hazards Centre, 1991.

London Hazards Centre. *Fluorescent Lighting: A health hazard overhead*. London: London Hazards Centre, 1987.

London Hazards Centre. *Hard Labour: Stress, Ill Health and Hazardous Employment Practices*, London: London Hazards Centre Trust, 1994.

London Hazards Centre. *Protecting the Community: A Worker's Guide to Health and Safety in Europe*. London: London Hazards Centre, 1992.

London Hazards Centre. *Sick Building Syndrome: Causes, Effects and Control*. London: London Hazards Centre, 1990.

London Hazards Centre. *VDU Work and the Hazards to Health*. London: London Hazards Centre, 1993.

McCann, M. *Artist Beware*. New York NY: Lyons & Burford, 1992.

McGuire, Scarlett. *Best Companies for Women*. London: Pandora Press, 1992.

Mackay, C. 'Visual Display Units', *Practitioner*, 1989; 233:1496–8.

McPartland, P. *Promoting Health in the Workplace*. New York NY: Harwood Academic, 1991.

Manser, W.W. 'Lead: A Review of the Recent Literature', *Journal of the Pakistan Medical Association*, 1989; 39:296–302.

Merletti, F., Heseltine, E., Saracci, L. *et al.* 'Target Organs for Carcinogenicity of Chemicals and Industrial Exposures in Humans: A Review of Results in the IARC Monographs on the Evaluation of the Carcinogenic Risk of Chemicals to Humans', *Cancer Research*, 1984; 44:2244–50.

National Committee for Clinical Laboratory Standards (NCCLS) 'Protection of Laboratory Workers from Infectious Disease Transmitted by Blood and Tissue; Proposed Guideline', NCCLS Document M29–P, 1987.

National Council for One Parent Families. *Returning to Work: A Guide for Lone Parents*. London: National Council for One Parent Families, 1994.

National Council on Radiation Protection and Measurements. *Recommendations of the National Council on Radiation Protection and Measurements*. Bethesda MD: National Council on Radiation Protection and Measurements, 1993.

National Out of School Alliance. *Guidelines for Good Practice for Out of School Care Schemes*. Reading: Reading Borough Council, 1988.

Needleman, H.L. 'What Can the Study of Lead Teach Us About Other Toxicants?', *Environmental Health Perspectives*, 1990; 86:183–9.

Needleman, H. and Bellinger, D. eds. *Prenatal Exposure to Toxicants: Developmental Consequences*. Baltimore: The Johns Hopkins University Press, 1994.

Nelson, K.E. and Sullivan-Bolyai, J.Z. 'Preventing Teratogenic Viral Infections in Hospital Employees: The Cases of Rubella, Cytomegalovirus and Varicella–Zoster Virus', *State of the Art Review, Occupational Medicine*, 1987, 2:3:471–98.

Nelson, L., Kenen, R. and Klitzman, S. *Turning Things Around: A Women's Occupational and Environmental Health Resource Guide*. Washington DC: The National Women's Health Network, 1990.

New Ways to Work. *Balanced Lives: Changing Work Patterns for Men*. London: New Ways to Work, 1995.

Newell, S. *The Healthy Organization: Fairness, Ethics and Effective Management*. London: Routledge, 1995.

Nurminen, T., Jusa, S. Ilmarinen, J. and Kurppa, K. 'Physical Work Load, Fetal Development and Course of Pregnancy', *Scandinavian Journal of Work and Environmental Health*, 1989; 15:404–14.

Nurminen, T. 'Shift Work, Fetal Development and Course of Pregnancy', *Scandinavian Journal of Work and Environmental Health*, 1989; 15; 395–403.

Occupational Safety and Health Information Group. *Health and Safety Journals: What's Where*. Cambridge: Occupational Safety and Health Information Group, 1993.

Olsen, J., Hemminki, K. *et al.* 'Low Birthweight, Congenital Malformations, and Spontaneous Abortions Among Dry-Cleaning Workers in Scandinavia', *Scandinavian Journal of Work and Environmental Health*, 1990; 16:163–8.

Palmer, C. *Maternity Rights*. London: Legal Action Group and Maternity Alliance, 1997.

Pantry, S. *Dealing with Aggression and Violence in Your Workplace*. London: Library Association, 1996.

Participatory Research Group. *Short Circuit: Women in the Automated Office*, Toronto, Ontario, 1985.

Paul, M., ed. *Occupational and Environmental Reproductive Hazards: A Guide for Clinicians*. Baltimore: Williams and Wilkins, 1992.

Paul, M. and Himmelstein, J. 'Reproductive Hazards in the Workplace: What the Practitioner Needs to Know About Chemical Exposures', *Obstetrics and Gynecology*, 1988; 71 (6 Part 1):921–8.

Pelmear, P. 'Low Frequency Noise and Vibrations: Role of Government in Occupational Disease', *Semin Perinatol*, 1990; 14:322–8.

Persaud, T. 'The Pregnant Woman in the Workplace: Potential Embryopathic Risks', *Anatomischer Anzeiger*, 1990; 170:295–300.

Ratcliffe, J.M., Schrader, S.M., Clapp, D.E. *et al.* 'Semen Quality in Workers Exposed to 2–Ethoxyethanol', *British Journal of Industrial Medicine*, 1989; 46:399–406.

Ricci, E. 'Reproductive Hazards in the Workplace', *NAACOGS Clinical Issues in Perinatal Womens Health Nursing*, 1990; 1:226–39.

Robinson, J. *Toil and Toxics: Workplace Struggles and Political Strategies for Occupational Health*. Berkeley CA: University of California , 1991.

Rogan, W.J. 'Breastfeeding in the Workplace', *Occupational Medicine*, 1986; 411–13.

Savitz, D.A., John, E.M. and Kieckner, R.C. 'Magnetic Field Exposure From Electric Appliances and Childhood Cancer', *American Journal of Epidemiology*, 1990; 131:763–73.

Schrag, S.D. and Dixon, R. 'Occupational Exposures Associated with Male Reproductive Dysfunction', *Annual Review of Pharmacology and Toxicology*, 1985; 25:567–92.

Schrag, S.D. and Dixon, R. *WOHRC News*, vol. 8, no. 2.1, New York: Women's Occupational Health Resource Centre, School of Public Health, Columbia University.

Scott, A.J. and LaDou, J. 'Shiftwork: Effects on Sleep and Health with Recommendations for Medical Surveillance and Screening', *Occupational Medicine*, 1990; 5:273–99

Selevan, S.B., Lindbohn, M.L. *et al.* 'A Study of Occupational Exposure to Anti-Neoplastic Drugs and Fetal Loss in Nurses', *New England Journal of Medicine*, 1985, 313:1173–7.

Sheehan, H. and Wedeen, R. *Toxic Circles: Environmental Hazards from the Workplace into the Community*. New Brunswick NJ: Rutgers, 1993.

Shepard, T. *Catalog of Teratogenic Agents*. (8th ed.) Baltimore: The Johns Hopkins University Press, 1995.

Shooter, R. *A Working Choice for Parents: The Case for a National Childcare Policy*. London: Working for Childcare, 1991.

Shortridge, L.A. 'Advances in the Assessment of the Effect of Environmental and Occupational Toxins on Reproduction', *Journal of Perinatal and Neonatal Nursing*, 1990; 3:1–11.

Singer, E. *Risky Business: Genetic Testing and Exclusionary Practices in the Hazardous Workplace*. New York: Cambridge, 1991.

Sorahan, T. and Waterhouse, J.A.H. 'Cancer Incidence and Cancer Mortality in a Cohort of Semiconductor Workers', *British Journal Industrial Medicine*, 1985; 42:546–50.

Stein, Z. and Hatch, M., eds. 'Reproductive Problems in the Workplace, State of the Art Reviews', *Occupational Medicine*, 1986; 3.

Stellman, J. and Henifin, M.S. *Office Work Can Be Dangerous to Your Health* (revised and updated edition). New York: Fawcett Crest, 1989.

Stoker L. *Women Returners' Guide*. London: Bloomsbury, 1991.

Stranks, J. *Human Factors and Safety*. London: Pitman, 1994.

Stucker, I. and Caillard, J. 'Risk of Spontaneous Abortion Among Nurses Handling Antineoplastic Drugs', *Scandinavian Journal of Work and Environmental Health*, 1990; 16:102–7.

Suess, M.J., ed. *Nonionizing Radiation Protection*, (Series No. 10) Copenhagen: World Health Organization, Regional Office for Europe, 1982.

Symposium on Women's Occupational Health. *Women and Health*. Symposium on Women's Occupational Health, 18:3, 1992.

Teichman, R.F., Fallon, L.F. Jr. and Brandt-Rauf, P.W. 'Health Effects on Workers in the Pharmaceutical Industry: A Review', *Journal of Sociological Occupational Medicine*, 1988; 38:55–7.

Thomas, J.A. and Ballantyne, B. 'Occupational Reproductive Risks: Sources, Surveillance, and Testing', *Journal of Occupational Medicine*, 1990; 32:547–54.

Trades Union Congress. *TUC Guidelines on Sexual Harassment*. London: Trades Union Congress, 1991.

Trades Union Congress. *Women's Health at Risk*. London: Trades Union Congress, 1991.

Trades Union Congress. *Working Women: A TUC Handbook for All Trade Unionists*. London: Trades Union Congress, 1991.

Trades Union Congress. *Your Voice at Work: TUC Proposals for Rights to Representation at Work*. London: Trades Union Congress, 1995.

US Congress, Office of Technology Assessment. *Reproductive Health Hazards in the Workplace*, Washington DC, 1985. US Government Printing Office, OTA-BA-266.

US Department of Health and Human Services, *Shiftwork and Health*, Washington DC, 1986, DHHS pub. # (NIOSH) 76–203.

US Department of Health and Human Services, *Working in Hot Environments*, Washington DC, 1986 DHHS pub. # (NIOSH) 76–203.

Weaver, C. 'Toxics and Male Infertility', *Public Citizen*, 1986; 7.

Willetts, D. *Happy Families? Four Points to a Conservative Family Policy*. London: Centre for Policy Studies, 1991.

Williams, L.A. *Reproductive Health Hazards in the Workplace*. Philadelphia: Science Information Resource Center, 1988.

Williams, V., ed. *Babies in Daycare: An Examination of the Issues*. London: The Daycare Trust, 1989.

Wolff, M. 'Occupationally Derived Chemicals in Breast Milk', *American Journal of Industrial Medicine*, 1983; 4:259–81.

Woods, M. and Whitehead, J. in association with Lamplugh, D. *Working Alone: Surviving and Thriving*. London: Pitman, 1993.

Working Mothers' Association. *The Employer's Guide to Childcare*. 2nd ed. London: Working Mothers' Association, 1991.

World Health Organization. *Control Technology for the Formulation and Packing of Pesticides*. World Health Organization, 1992.

Working Mothers' Association. *The Working Mother's Handbook: A Practical Guide to the Alternatives in Childcare*. 6 ed. London: Working Mothers' Association, 1991.

World Health Organization. 'Visual Display Terminals and Workers' Health', World Health Organization, WHO Offset Publications, 1987:99:1–206.

Zakstein, K. *Occupational Exposure To Chemicals in Pregnancy. Maternal–Fetal Toxicology: A Clinician's Guide*. New York: Marcel Dekker, 1990.

Zielhuis, B.L. *Health Risks to Female Workers in Occupational Exposure to Chemical Agents*. Berlin: Springer–Verlag, 1984.

Glossary of Occupational Health Terms

acute effects – effects which are seen shortly after exposure to a toxic material, usually at a fairly high concentration.

animal studies – studies in which animals (e.g. rats, mice, hamsters) are exposed to chemicals for different periods of times and at different levels of exposure to see whether or not any develop cancer, or other diseases, over their lifetime at a significantly higher rate than a 'control' group of animals (a group that has not been exposed).

bacterial studies – a way of looking at millions of living organisms at one time by studying those simpler than mammals, as genetic material is the same in all living organisms.

bias – a condition that might invalidate study findings. Three common kinds of bias are selection, observation and recall bias.

selection bias – volunteers who participate in a study may not be typical of the population they are supposed to represent.

recall bias – persons who have suffered reproductive harm are likely to recall more detailed and accurate information about prior events than those who have had no reproductive problems.

observation bias – the researcher is influenced by knowledge of whether a subject in the study is a member of the case or control group.

birth defect – an abnormality in an infant that may be seen at birth, or noticed at some point later in development.

breastmilk pollution – substances toxic to human health enter the mother's milk, primarily through the food chain, and are transmitted to the nursing infant.

carcinogen – a substance capable of causing cancer.

case-control design – a study based on a comparison of individuals who are suffering from a disease with a matched group of individuals who do not have the disease.

chromosomes – rod-like structures containing the genes which are found in the cell nuclei.

chronic effects – health problems which appear a relatively long time after a person's *first* exposure.

cohort design – follows a group of people to determine who comes down with the disease being studied. Researcher looks for characteristics or exposures that differentiate between those who become ill and those who do not.

confounder – a factor that is associated with both the exposure *and* the outcome under study, making it difficult for the researcher to determine whether the effect is due to the exposure or to the confounding factor.

congenital – present at birth.

cytogenetic studies – observation of cells under an electron microscope to detect chromosomal abnormalities that might be a result of exposure to a mutagen.

dose-response assessment – a process that determines the relationship between the magnitude of human exposure and the probability of human health effects.

embryo – from conception through the twelfth week of pregnancy.

embryotoxic – a substance that is toxic to the embryo.

engineering control – changing equipment design and processes in order to reduce the amount of hazardous substances to which workers are exposed.

epidemiology – the study of causes, patterns and distribution of diseases in human populations.

foetal protection policy – an occupational health policy in the United States that prevents fertile and/or pregnant women from holding certain jobs that are deemed to pose a hazard to future offspring.

foetus – an unborn child from twelve weeks until birth.

gene – a unit of heredity comprising a segment of DNA.

genetic defect – an abnormality in the genetic material of cells (the genes or the chromosomes).

germ cell – the egg or sperm cell containing reproductive material which determines the characteristics that will be inherited by the young from its parents.

ionising radiation – X-rays, gamma rays, alpha and beta rays which release energy capable of causing ionisation of atoms or molecules in radiated tissue.

local effect – the action of a substance which occurs at the point of contact, e.g. a skin rash.

mutagen – a chemical substance or physical agent that can cause mutations (changes) in the genetic material of living cells.

mutation – a change (usually harmful) in the genetic material of a cell. When it occurs in the germ cell, the mutation can be passed on to future generations.

neonatal – affecting or relating to the newly born.

NOEL (No Observed Effects Level) – the maximum level of exposure that appears to produce no harmful effects for most individuals.

non-ionising radiation – radiation that is lower in energy than X-rays, e.g. visible light, infrared radiation, microwaves and radio waves.

personal protective equipment (PPE) – clothing or equipment, such as gloves, hearing protectors, and gas and dust masks worn by the person and designed to reduce exposure to potentially hazardous substances.

placental barrier – the border to the placenta (the organ connecting the embryo to the mother's uterus). Some toxins that the mother is exposed to can cross this barrier.

power of a study – the ability of the study design and sample size to detect a real association between exposure and outcome.

radiation – when X-rays, gamma rays, alpha or beta particles pass through matter.

somatic cells – all the cells of the body, other than the germ cells.

spermatogenesis – the development of sperm cells.

synergistic effect – an interaction between exposure to two or more substances or agents. This combination causes a greater effect than exposure to either by itself.

systemic effect – when the action of the chemical or substance does not occur at the point of contact, but travels through the system and damages another organ e.g. vinyl chloride can enter through the lungs but can cause cancer of the liver.

teratogen – a substance or agent that interferes with the development of the embryo or foetus during gestation.

threshold limit values – ACGIH guidelines which establish the maximum level of specific chemicals to which workers may be exposed without experiencing a harmful effect.

tight building syndrome – (or closed building syndrome) the build-up of indoor pollution in buildings having inadequate ventilation systems which affect the health of workers, such as headaches, allergies, eye irritation, breathing difficulties.

tort – a wrongful act for which an employer can be judged legally liable in the United States.

toxic – damage by a substance to cells, tissues and organs.

Index